THE PIGKEEPER'S GUIDE

THE PIGKEEPER'S GUIDE

Peter Mitchelmore

DAVID & CHARLES
Newton Abbot   London   North Pomfret (Vt)

British Library Cataloguing in Publication Data
Mitchelmore, Peter
 The pigkeeper's guide.
 1. Swine
 I. Title
 636.4'083   SF 395
 ISBN 0-7153-7995-X

© Peter Mitchelmore 1981

All rights reserved. No part of this publication may be reproduced, stored in a retrieval system, or transmitted, in any form or by any means, electronic, mechanical, photocopying, recording or otherwise, without the prior permission of David & Charles (Publishers) Limited

Filmset in Monophoto Baskerville
by Latimer Trend & Company Ltd Plymouth
and printed in Great Britain
by Biddles Limited, Guildford, Surrey
for David & Charles (Publishers) Limited
Brunel House   Newton Abbot   Devon

Published in the United States of America
by David & Charles Inc
North Pomfret   Vermont 05053   USA

Contents

	Glossary	7
1	Beginning Pigkeeping	9
2	The Pig	12
3	Breeds	17
4	Housing	26
5	Fencing	42
6	Feeding	45
7	General Routine	51
8	Equipment	53
9	The Boar	54
10	Breeding	58
11	Artificial Insemination	65
12	Farrowing	67
13	Weaning	85
14	Modern Pigkeeping	87
15	Bedding	89
16	Dung—How Much?	93
17	Teeth Cutting	95
18	Identification	96
19	Castration	98
20	Ringing	101

CONTENTS

21	Fighting and Mixing Pigs	103
22	Artificial Rearing	107
23	Diseases	109
24	Poisons	124

Glossary

ADAS Agricultural Development Advisory Service, a ministry body.
AI Artificial insemination.
ark, pig ark Low, wood-floored pig house set in a field.
bacon pig Selected for its length and leanness, and slaughtered for bacon at 86 to 95kg (190 to 210lb).
boar Male pig not castrated, used for mating with female breeding stock.
creep A place for piglets to lie beside their mothers, and to have a creep feed.
creep feed Food given to piglets while still in the creep.
cobs, pig cobs Pigs' main diet. Like pig nuts. Milled grain with added protein.
controlled environment house Pig house with controlled temperature, controlled air flow and subdued lighting.
cutter In between a pork pig and a bacon pig in size; dead weight 54 to 77kg (119 to 170lb).
empty or barren sow A sow that is not pregnant, most likely being fattened up to slaughter.
entire Non-castrated male.
farrowing Giving birth to a litter of pigs.
farrowing crate Crate provided for the sow to farrow in.
gestation period For a sow, 3 months, 3 weeks and 3 days; the sow's normal temperature during gestation is 4·2°C (39·5°F), taken by inserting a thermometer in the pig's rectum.
gilt Female pig having a litter for the first time.
heat period Also known as 'boaring' and 'in season'; the period when an adult female will mate with the boar; the female comes in season every twenty-one days provided it is not pregnant and not with a litter.
heavy hog Heavier than a bacon pig, 77 to 100kg (170 to 220lb); used for bacon, with the fat cut away and used in sausages, etc.
hog Castrated male.
maiden gilt Female yet to have her first litter.
mange wash Liquid mixed with water to kill mites, lice, etc.

GLOSSARY

MLC Meat and Livestock Commission, a ministry body.
multisuckling Sows and litters put together in one house.
nestlebird, nestledraft Runt, smallest pig in a litter.
notifiable disease Under official control by the Ministry of Agriculture, who must be informed if a case is confirmed.
pig board Piece of plywood to guide pigs into sty, for instance.
piglet Baby pig before it has been weaned.
pork pig Pig reared for pork, slaughtered at about 45kg (100lb) dead weight.
prick ear Type of pig with ears rising above the head.
ringing Inserting a ring into the pig's snout to prevent it from rooting up the ground.
runt Smallest pig in a litter.
scour Virulent form of diarrhoea common to pigs.
serving The mating of a sow with a boar.
serving crate Crate equipped to enable a sow to be served easily.
sow Female pig used for breeding and rearing piglets.
stalls Used for sows; individual steel-bar compartments in rows.
store Pig of either sex from weaner to pork weight.
tethers Used on sows, kept by a tether around their neck or girth.
weaner A pig that has just been weaned from its mother's milk; weaners are sold at 18 to 27kg (40 to 60lb).

1
Beginning Pigkeeping

In the ever-growing, rushing world of science and high technology, with its mental and physical pressures, it is little wonder that more and more people today are finding enjoyment, happiness and fulfilment in being what I would call close to the land or close to nature in their own way. Large numbers of people now own a few acres of land (or not even as much as that), perhaps growing vegetables, or keeping animals, or both. The demand for this type of property is increasing all the time, and those few acres added to the dwelling command a premium.

It is to those of you lucky enough to have this sort of accommodation that this book is directed, intended as a guide for the beginner who wants to keep pigs.

Pigs are simple, lovable creatures but, like all animals, they demand a certain amount of attention if they are to be contented and profitable.

You should enjoy looking after your pigs, give freely of your labours and take a genuine interest in their welfare, because when pigkeeping becomes a grudging chore, you and the pigs will suffer. Then I would suggest that it would be better if you did not keep them at all.

One thing in the pigs' favour, to keep your interest, is that they are, in farming terms, a 'quick turnover'. From birth a pig can be a pork pig in as little as twelve weeks, but on average a little more; also the gestation period for sows is just under four months, so you can see things move along quite quickly with pigs.

There are a number of ways you can make a start. Pigs can be obtained quite easily from various sources. They can be purchased at your local market, a practice I am not very keen on as you do not always know the reason for them being sold; there is no point in 'buying trouble', unless of course you know and trust the person selling. Another source is from your local auctioneer, the Meat and Livestock Commission (MLC) or your local Agricultural Development Advisory Service (ADAS) officer, who will be only too pleased to put you in touch with some reputable pig breeders. Many good breeders advertise in local daily papers. My advice is to buy from the local small man who has a reputation for breeding good sound pigs.

Ministry organisations like MLC and ADAS do a magnificent job for our farming community, and have experts in every field of agriculture. Their help and advice are invaluable, always available at the end of a telephone, a service we sometimes take for granted in this age of cuts and economies. One wonders how long it will be before this valuable service will be cut out. Perhaps I have given the impression that these services are all free, but there is a small levy on all pigs slaughtered, although it is only a few pence per pig.

Ask and accept advice from people with experience, there is no substitute for it. Take advantage of any pig meetings or discussion groups. ADAS and MLC will be only too pleased to put you on their mailing list. One can never stop learning about pigs; I like to think I learn something new every day.

Observe and study your pigs. Jokes are often made about farmers leaning over a field gate idling their time away looking at their stock. In fact some of the best ideas are derived from this sort of activity. Beware of, or at least be sceptical of, agricultural representatives who come around confidently predicting that their new animal food will save you ten per cent on your food bill, or selling a disinfectant that is fifty per cent more effective, or an additive that will cure all ills.

If all these claims were true, bacon pigs would be going to

the factory soon after they were born, and every disease would be eliminated.

A few years ago a certain firm bought out a creep feed pellet with a particular flavour to it, which it claimed would revolutionise the creep feeding of piglets (still in the creep with their mothers). This pellet in fact does no better than any other firm would claim for their product.

Pigs and pig houses have a particular smell of their own—unlike cows, sheep or poultry. Someone not familiar with animal life would say they stink, and they would be right; but this is something that you get used to and take no notice of. I have sometimes gone out in the evening, wearing good clothes, and, coming home, have slipped into the farrowing house to check that the sows are comfortable prior to giving birth. On my going indoors and into the bedroom, my wife has said, 'You can smell where you have been.' To me this is no different from someone going into a odourless atmosphere after frying chips, or coming from a smoke-filled room; the smells then stick out like a sore thumb.

Never be complacent in any area of your pig enterprise, just because your pigs are doing well; don't try to cut corners by not scrubbing out the pens or disinfecting them when you should. Deal promptly with any problems regarding the health of your pigs.

2
The Pig

Years ago, when I was a lad, I used to keep pigs. I found myself with a little pig which was really what we would call a runt or nestle bird, being the smallest of a litter of twelve. Because he was small he was always being pushed around, and so he did not get as much to eat as he should have. He didn't look like a runt, he was just small; I did not really want him as I had no other pigs his size to put him with and no spare house to keep him in, so I gave him to a friend of mine whose parents kept a few cows on a smallholding. There he was fed on meal, scraps and skimmed milk. After a day or two, when he had settled down, he really began to grow, and as so often happens he became a real pet. They used to let him out and he would play like a dog, and on fine evenings they used to put a harness on him and take him up the road onto the green; while we played football the pig would run around, having a fine old time. When it was time to go home my friend would shout, 'Come on pig!' and he would come running, have his lead put back on, and down the road he would go.

Inevitably the time came when this pig was big enough to go to the bacon factory, and although it seems an awful thing to do to a favourite it is a situation that will always arise when a pig is originally kept and reared for meat, but ends up being a pet and one of the family.

So if you are contemplating rearing a pig or two for the freezer, make sure you buy gilts (young breeding females) not hogs (castrated males), so that if the worst comes to the worst you can keep them on and in due course breed from them. This gets you over the problem of not having the courage of your convictions when it comes to sending the pigs to the slaughterhouse.

Pigkeepers often feel especially tender-hearted towards their first pigs, but by the time the pigs have had little ones and they start growing up, they are not seen in quite the same light. You find it easier to accept that they have to go.

People still use pigs for finding truffles, tuberlike fungi that grow under the ground and which are eatable, with a nutty flavour, and much sought after for cooking. The pig, especially an older sow, has the ability to know whereabouts a truffle is under the ground, either by instinct or by smell. When trained by its owner, the pig will proceed to root for truffles; when it is evident that the pig has found one, the truffle is secured and man and beast move on to find another one, thus eliminating the necessity to dig up the whole wood to find them.

For thousands of years the wild pig has had to fend for itself. It is not really an aggressive animal—just sometimes anxious to defend its food, home or family, and anyone that it feels belongs to it. If you lie down in the middle of a group of fat pigs, or a group of sows, they will cheerfully chew you up and eat you, not through malice—it is just their natural way of rooting and eating. You read from time to time in the newspapers of someone getting mauled or even killed by pigs in a field. I doubt that pigs would do this intentionally, as would a lion or tiger; it is probably a question of someone walking across a field with pigs in it, seeing them approach, panicking, and most likely fainting—in which case the pigs will do just that—eat the person.

It is no joke being confronted by a savage pig which really means to have a go at you. This can happen sometimes with a sow who things she is protecting her young.

If you think there is any chance of this situation arising when handling piglets that may squeal, make sure the mother and any other mothers near are safe in their pens; make sure the mother cannot reach you, or take the piglets away out of sight and sound altogether. Never try to calm her down, scold or hit her, it will only make her worse; just ignore her and she will soon forget about it.

When endeavouring to break a fodder beet in half by bringing my foot down upon it, and making no impression except for nearly breaking my ankle, it brings home to me the power in the jaws of the pig, which with one bite will break it into bits.

When moving pigs, and particularly when handling or driving boars (Chapter 9), always use a pig board, galvanised iron, or a sheet of anything, not just for the safety angle, but because it is so much easier: if the pig can see daylight it will come straight on past you.

Pigs have always been either 'all pounds, or all pence', as the saying goes, partly because the pig population can be quickly increased and also because there is no effective meat marketing board such as the Milk Marketing Board to set a price that pigkeepers can all work to.

More than half the bacon consumed in this country is foreign. Foreign producers have been paid, up to now, handsome subsidies for their trouble, and home producers cannot possibly compete on equal terms; our canned meat industry is nearly nonexistent. Think of the extra jobs that would be created if we were self-supporting in all pig products. More abattoirs, more transport, canning factories, bacon factories, and more people to look after the pigs in the first place. Extra jobs would be created in the building industry, to provide more equipment and pig houses, plus welcome business for offshoot companies concerned with the industry. Millions of pounds could be saved on imported meat.

Pigs are like most animals in that they will adapt to almost any conditions they find themselves in. Not so much today, but in years gone by, the pig was a very important feature of the country scene. At harvest time, when the corn and straw had been gathered in, one or more sows would be let into the field and could live there for days on the grain that was left behind.

These sows would know the drill; they would be in the orchard or the meadow all summer, then out in the cornfield. When they had finished scouring there, they would be moved to where the potatoes and root crops had been grown and harvested, and would spend some time there rooting out what had been left behind. A low-roofed pig ark might be put in the field for them, but more times than not they would sleep under the hedge. These pigs were hardy and were happier outside than in a pen.

With the approach of Christmas, the apples in the orchards would have been picked up, enabling the sows to go back there to grub out any that were left in the grass.

Acorns also played an important part in the pig's diet. Indeed, sows would be served, and then 'turned away' into the woods to consume acorns, nuts and anything else that took their fancy. When it was time for a sow to farrow she would make up her bed in the wood and have her litter there. One day she would emerge with her piglets running to foot. There may have been only five or six that survived, but these piglets were free, gratis and for nothing; the farmer had not spent a penny on food whatsoever.

Farmers today will tell you that they haven't the time or cannot be bothered to mess around with only a few pigs like that, and in any case modern farming does not lend itself well to that sort of enterprise. There are not half the oak trees or woods any more and, alas, very few apple orchards either.

These are treasured memories for me; we can never again expect to see the traditional ways of raising pigs.

It is natural for pigs to snuggle up to each other as they would in the wild; but if they started to fall out, in the wild at least they could move off out of the way—not so when a group of sows are in a relatively small house. That is why I am sure the underdogs, the ones that are bullied, are really glad to be put into a farrowing crate when their turn comes, for a bit of peace of mind, knowing that they can lie out without having to be constantly on their guard against someone giving them a sly nip before they have a chance to get out of the way. From this point of view, there is much in favour of the crates, stalls and tethers.

Over the years, particularly the last ten years, pig herds have got larger and larger. In 1970 there were 60,000 pig-farming units, while the total for 1980 is expected to be down to 25,000. Not that the pig population has decreased by any appreciable amount; it is just that the herds of today have increased in size, on average threefold; people with the odd sow or fattening pigs are not included in this census.

A person keeping less than a hundred sows has found it increasingly difficult over the years to make enough profit for a reasonable income to live on; so herds have either been given up altogether or doubled or trebled in size. Like all agricultural enterprises, pigkeeping has to expand to stand still, as it were. Another factor has been the increasing involvement of the big feed-companies, sinking huge amounts of cash into very large pig enterprises.

On the other hand, it is pleasing to hear that animal smallholdings are now on the increase, particularly with the rare breeds—not only with pigs, but with all kinds of animals and birds.

3
Breeds

There are perhaps a dozen distinct breeds of pig in England being used on a commercial scale to fulfil a specific need in a given area. As long as the buyers of the meat are happy (the butchers, the supermarkets, etc) then the pig farmer will be very reluctant to change his system and particularly the breed of pig he is using. This is why there are pockets of any given breed of pig all over the country. Quite a few breeds are more or less kept going by people who have bred a certain breed of pig for years and just like to see a few sows about the place, perhaps for sentimental reasons, if nothing else.

Some breeds are now extinct, like the Small White, which was a very fat animal with a short face and was originally imported from China; its ancestor was the Asian wild boar. Many breeds originated from several different countries and were probably crossed with our British wild pig. It is said that the domestic pig descended from the wild pig of Eastern Asia and the wild pig of Europe. In the early days the British pig was a yellowy brown colour with dark markings. The Tamworth still keeps its brown colour given to it by the imported Red Barbdan.

I am not going to suggest which breed of pig you purchase. I think it is a matter of personal choice, and in truth I do not think there is a lot of difference between them. However, if you are contemplating selling weaners that will end up in the big fattening units you will find them easier to sell if they are white, as it is considered that whites grade better when slaughtered, in our complicated grading system. By white I mean the modern Large White, Landrace and the Welsh types.

There is no particular breed that is best suited to running outside, as all pigs would like to be able to go out onto grass when they felt like it. It is the type of housing that is the governing factor. I know a farmer who keeps his sows out all the year round, and their only protection from the weather is half-round galvanised sheds, open both ends. They farrow there too. These sows are especially bred for the purpose, with a proportion of Saddleback blood in them. It would be inadvisable to try this system with the modern white breeds.

Here is some information about a number of the breeds.

The Berkshire

Not many of this breed about at the moment. It was in the Thames Valley that the Berkshire was first recognised, its ancestors probably being of Asian origin or from China in the late eighteenth and early nineteenth centuries. It is an early-maturing pig with a short head and short compact body. At one time it used to be reddish brown in colour with darker patches, but the Berkshire today is almost black with white feet and white on the end of its tail and snout, the white bits hardly noticeable; it also has prick ears (ears rising above the head).

The Berkshire used to be *the* pig for the pork trade, ideally suited because of its early maturity, and at one time would have taken all the prizes in the pork carcass classes. It has since been exported to many parts of the world.

In the USA the Berkshire is said to have contributed to the formation of the Poland China breed. It does well out of doors and is of good temperament.

British Lop Ear

The sort of pig people kept in their back yards, this is a very localised pig, closely related to the Landrace, but not quite so 'fine'. It has not been subjected to such intensive breeding as other kinds of pig. They used to be known as the Long White Lop Ear, originating from south-western England. They still play a useful part in the pig world and are quite hardy.

British Saddleback

My favourite breed of pig, I think probably because it was the first pig I came into contact with when I was young. This is not a very old breed, first coming into existence early in the nineteenth century with the amalgamation of the Wessex, from Dorset, and the Essex, from East Anglia; both breeds have a white saddle over the shoulders, hence the name; other than this they are all black except that some have white feet and tail. As with all pig breeds there is a very wide variation in type and colour.

Saddlebacks are often used in the makeup of a particular strain required for outdoor rearing, as they are hardy, good mothers and produce good strong litters. They were very popular thirty to forty years ago, until someone decided we must all have long, lean white pigs; then their numbers dropped. It is a fact that for someone producing weaners it is easier to sell white pigs rather than 'blue', the name given to white weaners with a spot or two of black on them. Nevertheless saddlebacks are still a popular breed, with classes for them in most big shows.

Chester White

As the name suggests, the Chester White originated in the Chester country of Pennsylvania. It came into being about 1920 by crossing three breeds, the Lincolnshire, the Cumberland, and the Yorkshire or Large White. The Lincolnshire curly coat and the Cumberland are now extinct— sad to think that a lot of these old breeds are lost for ever. The Chester White is a large breed and over the last hundred years or so has got bigger still, with mature boars weighing 450kg (900lb) or more. They are always white, with some types having tiny blue spots, like freckles on the skin. They produce a lean carcase of high quality, are very good mothers and produce large litters.

Durocs

A well-established American breed, there are now a few herds in this country. Durocs' frozen semen is available from Canada, at something like six times the price of our English breeds' semen. They are black pigs with funny little ears that seem to go halfway and then fold forwards. They are hard, sturdy pigs, docile, easy to manage and not clumsy. Durocs are highly prolific, often having fifteen to seventeen piglets, but not always of good size. They are seen in this country as pigs with important new bloodlines to improve our own British breeds. A Duroc boar put to a white breed will produce white pigs.

I was very interested to hear of the Iron Age Pig (as it was christened), bred specifically for the television series reconstruction of an Iron Age settlement; half a dozen young couples had volunteered to live for one year as near as possible to an Iron Age existence. The Iron Age Pig was derived from mating a Tamworth sow with a wild boar from London Zoo. The offspring was considered too wild for the project, but further inbreeding produced a pig that was suitable.

There is at the moment much interest being shown in foreign breeds—not in the pigs themselves (imports of pigs are generally forbidden) but in obtaining frozen semen for a particular breed over here, with a view to possibly improving the bloodlines of our own breeds, such as Durocs.

Gloucester Old Spot

A very local breed, the Gloucester was for many years not seen outside the Berkeley Vale, until a breed society was formed in

1914; in recent years the breed has spread all over England. Its attraction is probably the pleasant markings it has, all white with black spots. It is a real old-fashioned pig, a 'deep' pig, a 'piggy' pig. It was probably known first as the 'orchard pig' because it used to be kept in the many orchards in the Severn Valley, living on windfalls; it is still a good grazing breed. Gloucesters are perfectly hardy and are quite good mothers. Pig breeders in recent years have tended to breed out the spots, but people selling to self-sufficient types like to keep the spots; the nicer-looking spotted pigs sell better to people who only want to keep one or two sows.

The Hampshire

The Hampshire is an American breed recently introduced into this country with great success and used mainly for crossing purposes with our own breeds. It is arousing considerable interest throughout the world. It is one of the newer breeds being tried out in our quest for a better carcase. It provides one of the finest carcases with exceptionally high lean meat content and low fat.

The Hampshire resembles the Saddleback in colour with the white saddle and its white feet. The pigs' ears are short, small and pricked, they carry themselves well, being good on their feet, the breed is docile, and they make good mothers and milk well.

Landrace

As the name suggests, this is a native breed, in this instance native to Sweden, Norway and Denmark principally. Of all the breeds, the Landrace has to my mind been bred out of all relation to its true nature. European pig breeders have much to answer for; they have taken their breeding programmes to extremes in their efforts to breed a pig with fine shoulders, long lean body and wide backside. These features were all the rage some thirty to forty years ago when anything and everything was kept for breeding purposes (because of the demand); consequently many Landrace pigs were somewhat deformed, with bad backs, and legs too weak to walk properly on. Fortunately most of the defective specimens are gone today, but there are still defects in some types of this breed. However, there are some very good strains to be found. If you are contemplating the purchase of the Landrace, be on your guard against the weaknesses at the back end, and try to buy from someone with a good reputation for the breed. The pigs are all white with lop ears and long fine-boned snouts, are bred specifically for the bacon market and have an international reputation.

Large Black

Ancestors of the Large Black were said to have come from East Anglia and the South West. They are large pigs, when mature reaching 500kg (1,100lb), all black with long lop ears reaching the end of their snout. They have been exported to many parts of the world, especially hotter countries, where they do not get sunburnt, but being black I would have thought they attracted the heat more. They are pleasant, docile animals, very hardy, and have a reputation for doing well on poor rations. Perhaps not so prolific as some breeds, but they are good mothers.

Large White

There are more Large Whites in Britain than any other breed, and they have been exported to many countries, including the USA, where the breed is known as the Yorkshire. It evolved from Yorkshire and is said to have a little Chinese or Neapolitan blood in its ancestry. It is perhaps our largest pig, with mature boars weighing up to 510kg (1,122lb). It is essentially a prick-ear, is all white and quite long. As I have said before, different strains of the same breed vary enormously, and in this case some have real turned-up noses and others have straight ones. They are good multipurpose animals, being used for all aspects of pig production, and possessing a high proportion of lean meat. Large Whites are prolific breeders and wean their pigs well; not our hardiest breed, but will tolerate field conditions quite happily. Crossing the Large White with the Landrace is a common practice, aiming for the best of both breeds.

Middle White

The Middle White came into existence by crossing the Large White with the now extinct Small White. It enjoyed a period of popularity as recently as the first half of this century. Now it is quite rare and is in danger of extinction. It is all white, with short legs, and when heavy with pig the sow's udders will touch the floor. The pig has prick ears, very heavy jowls, and a short, turned-up nose which puts me in mind of a cross between a boxer dog and a pekingese. Very suitable for the pork trade, due to its early maturity.

Poland China

A different looking pig, the Poland China is all black except for its four white socks, white nose and tail, and has semi-lop ears and a medium snout. Once known as the Warren County Hog, so called because it was first raised in Warren County, USA, from an assortment of breeds including the China Pig, from Philadelphia, and the Berkshire. Poland Chinas are large pigs which grow fast with a high percentage of lean meat; never very popular in this country.

Tamworth

The Tamworth, with its long straight snout, is the pig which has most retained the look of its ancestors. It descended from the Red Barbadan or Axford pig and the Old British pig, which was a descendant of the European wild boar. Tamworths are a lovely browny red colour which derived from the Axford, and to see a litter of these little ginger pigs is a sight to behold. In England the Tamworth was widely used for the pork trade, especially in the Midlands.

The Tamworth is quite at home in hotter climates, as it does not sunburn like the white breeds, and has been exported widely. In North America it is used for bacon production. More recently in England Tamworths have grown quite popular with smallholders, with their long noses eminently suitable for running outside, reclaiming rough pastureland and scrub. They always seem to me to have a funny grin; I think it is their eyes. The sows are quite aimiable and good mothers, being gentle with their piglets. The breed can survive better on poorer food and will rough it better than most.

Welsh

The Welsh took off at about the same time as the Landrace became popular, again closely related. They have spread throughout the British Isles and are perhaps in between the Landrace and the British Lop, again long white pigs with lop ears. It would take an expert to distinguish between the Landrace and the Welsh in appearance.

Alas the Cumberland pig is no longer with us. As recently as 1960 it became apparent that there was only one Cumberland sow, called Sally, left and no boar of this breed to mate her to; so it was that the Cumberland pig passed into extinction. It hardly seems possible that less than forty years ago this breed was famous for its hams, bacon and sausages; even today these products still carry the Cumberland name, even though there are no pure Cumberlands left.

These days we are more alert to animals becoming extinct, and are made more aware of these situations by a body called

BREEDS

the Rare Breeds Survival Trust, who do a tremendous job in keeping up our interest in the rare breeds. They produce a monthly journal and have an annual rare breeds show and sale at the Royal Show Ground at Stoneleigh, Warwickshire.

4
Housing

Pigs have been and still are being kept in every conceivable type of building, with every type of material being used. Concrete blocks and concrete for the floors have in the past been used extensively. Today there are so many materials and different house-designs being used that it is hard to know where to begin.

As this book is intended for people who are contemplating pigkeeping for the first time, or for those who have one or two pigs already, I think it would be best to concentrate on the types of building you may already have; buying purpose-built pig housing today is a very expensive business and could not be economically justified for the odd sow or fattening pig.

Most people have a shed or building with their property that could be converted to pigkeeping quite easily and cheaply. Any old shed with a galvanised roof and thick damp walls can be made quite acceptable for pigs, and it will not cost you the earth. Three- or four-inch concrete blocks can be used to make an inside wall around the area where the pig will lie, leaving a two-inch cavity, or the pumice blocks which are common these days can be used; these are much warmer and lighter but a little more expensive to buy. Battens can be nailed to the wall and any kind of boarding can be nailed to this to form a cavity effect. Have the walls three or four feet high, a little higher for a sow. If the house is no wider than four or five feet, an old door or some planks can be put across to form a low warm place for the pigs to lie, and perhaps a few bales of straw could be kept up above. Only the part where you intend the pigs to lie

HOUSING

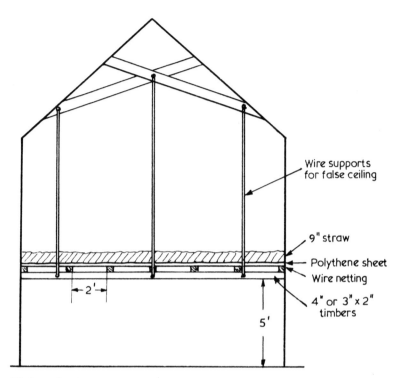

High-roofed house modified for pigs. Wire supports hold up ceiling of 3–4in timbers, covered with a layer of polythene (or fertiliser bags) and 9in of straw

need be done in this way; it is suprising how little room they need just to lie in. For a couple of pigs fattening to pork weight, an area as little as three by four feet would suffice, but an area twice as big as this would be needed for them to walk around and dung in. So many people keep one or two fattening pigs in a huge house with a little straw in one corner; if only they would make a box-type affair in the corner for the pigs to go in, they would be so much happier and grow a lot faster, not having to use quite so much of their food keeping themselves warm.

If you intend keeping several pigs together in a bigger house, the same principle can be used with just a false roof—pieces of

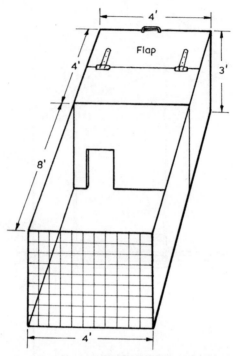

Kennel house for a litter of pigs already weaned. It is placed inside a bigger house. For outside use, add a double skin around the lying area and a waterproof top

four-by-two timber fixed across the house or even hung from the roof trestles with wire, about two feet apart. Lay over this wire netting, then a polythene sheet or old fertiliser bags, etc. On top of this sprinkle six to nine inches of straw, shavings, hay or any other such material, thus keeping your pigs warmer in winter and cooler in summer. This construction will not last for ever, but is cheap and effective compared with a permanent loft-type floor with floorboards where straw, etc, could be kept.

If your roof is low enough and has no insulation, nail up some ceiling boards—chipboard, hardboard or plywood—on the underside of the wood carrying the roof; this should be done with care, making sure that no rats or mice can get between the two and make their home there. Polystyrene is a marvellous insulator, but I would not recommend its use in the roof unless you can guarantee that no vermin can get near it. I have been into pig houses where the polystyrene has been literally chewed to bits, rendering it quite useless. Even a polythene sheet pinned up can make a terrific difference to the temperature, again warmer in winter and cooler in summer—just as people put polythene over their windows in winter—and of course it stops condensation. As an alternative, an idea that suited us for one of our sheds was to lay three or four layers of fertiliser bags on top of the existing galvanised roof, then nail on another complete galvanised roof; this did not in fact cost us anything (except for the nails) as we had plenty of spare galvanised iron. Galvanise can always be picked up quite cheaply at farm sales. Using this method we had no condensation, the house was much warmer, and no vermin could get between the two layers.

Pens can be built inside big buildings using concrete blocks, which are probably the cheapest material and very durable. There are several flatboard-type materials on the market, made of fibreglass, plastic, etc; these are very good for the pen divisions but are expensive. High roofs don't make for warm buildings; so, again as suggested, cover the lying area for warmth.

Hen battery houses are ideal for keeping pigs in, with the right sort of light and usually good insulation and ventilation; it is really only a question of how to design the inside.

You may have a building that can be used partly for pigs. The lying area could be in the house and a hole made in the wall for the run and dunging area outside. Heavy-duty weld mesh, the sort used in reinforced concrete in the building industry, is ideal for making outside runs for pigs, with very little extra support needed other than a few small stakes in the ground and pegs in the earth to keep the wire down. It may be possible to use part of a bigger house for sleeping quarters for pigs that are running out on grass, if the house is adjacent to the field. When this is not possible, a pig ark out in the field

HOUSING

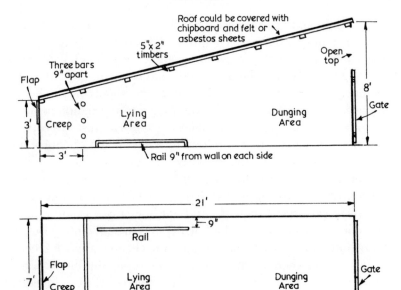

Multi-suckling house for four sows and their litters. It is divided into a 3ft wide creep, lying area (with rail 9in out from wall on both sides) and dunging area. The creep is separated from lying area by three bars 9in apart. The floor has an overall fall of about 3½in. The roof is 8ft high at the dunging end, sloping down to 3ft at the creep end. It rests on 5in by 2in timbers, and can be covered with chipboard and felt or asbestos sheets. The creep end has a flap, the dunging end a gate or door and is open above

would do very nicely; most of these have wooden floors and low roofs for warmth, and being low are not too much of an eyesore. There are plenty of firms who will be only too pleased to sell you a new ark, and they can be purchased without too much trouble—keep your eye on the local papers, as arks are always coming up in farm sales. To be on the safe side try to keep the purchased pig ark away from your or anyone else's pigs for a month or so, to prevent the continuance of any disease it may contain.

The monthly journal *Pig Farming* is excellent reading for

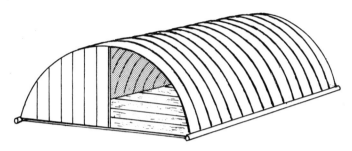

Ark for pigs in the field

anyone interested in pigs, and you can find every conceivable type of housing and information in it.

The building of the pens should be gone into thoroughly, and a great deal of thought put into it, as once the pens have been laid out and built, it is unlikely that they will be changed for a considerable time; it is important to get pens right at the outset.

Floors

Wooden floors are used for outside pig arks, or even the earth serves as a floor with bedding. Slurry systems must have all or part slatted floors with large channels underneath to take the slurry away, but keeping pigs on a small scale would not warrant the extra work and expense. Concrete floors are the most durable and when laid correctly are very satisfactory. A hard-soil base is needed to support a good concrete floor, the better the base the less concrete needed. A minimum of four inches of concrete is required if vehicles or tractors will be driving over it; this will not want to be as smooth as the lying area. A finish with a wood straight-edge is very good, or if too rough a soft long-handle brush run over it very lightly does a good job.

Before the floor is laid, there may be a lot of fill needed to bring it up to the required height. Any hard material will do,

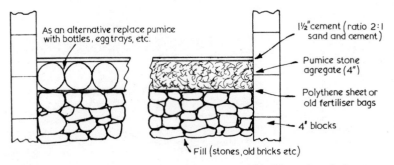

Housing: flooring bottom to top—fill (stones, old bricks etc), polythene or fertiliser bags, pumice stone aggregate, one-and-a-half inch skim of two parts sand to one part cement. Old bottles or egg trays can be used instead of pumice stone

like old brickwork, blocks or large stones. It is surprising how much hardcore can be picked up from a stony piece of ground, thus doing two jobs at once. The higher the fills, the less likelihood of damp rising. Before concreting, lay a polythene sheet over the area that will be the pigs' bed, or old fertiliser bags, which will eliminate any rising damp and keep the floor warmer.

To insulate the floor further, a variety of materials can be used; these can be placed on top of the polythene sheet. Egg trays are frequently used, and corrugated asbestos sheets work on the same principle, creating layers of air underneath the floor, making the floor warmer and insulating it as well. A layer of bottles over the floor makes an excellent insulation. You can also purchase special insulation blocks for this purpose—some firms are finally waking up to the fact that there is a market for aggregate materials. There is one, a cinder-type material which comes in approximately half-inch-diameter balls, which is mixed with cement to produce a very light and porous base, then finished off with less than a half-inch of cement and sand. Another material which is light and porous is pumice stone, imported from Italy, which can be used in the same way as the cinder. A good half-inch thickness is sufficient over the highest parts (top of egg trays, etc), with a quarter- to

Berkshire owned by W.D. and U.C. Mills, Crediton, Devon

Gloucester Old Spot (*National Pig Breeders' Association*)

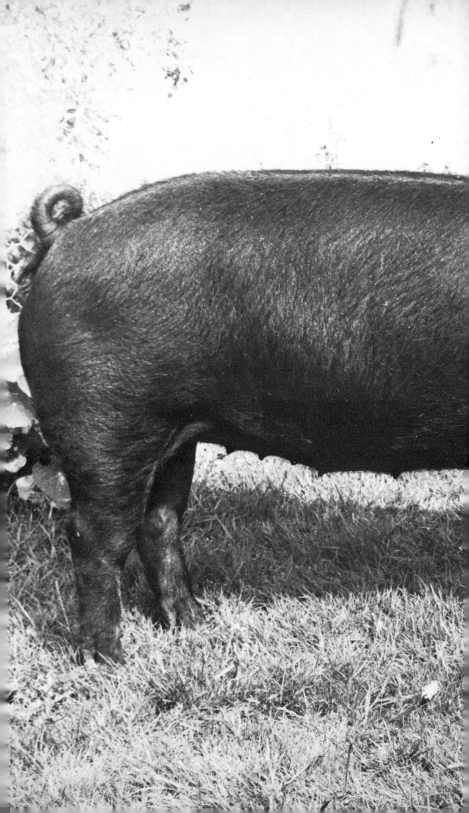

Hampshire sow
(*National Pig Breeders' Association*)

Hampshire crossed with Landrace gilt

Large White crossed with Landrace gilt, about to be scrubbed a week before she is due to farrow

Middle White owned by W.D. and U.C. Mills

Landrace sow (*National Pig Breeders' Association*)

Mature Large White boar

Tamworth owned by Mr and Mrs J. Buckley

Kennel house for weaners

Zig-zag fattening house at Stoneleigh. The lying area of another section is behind the wall on the left

Fattening house, also at Stoneleigh, showing half-slatted floor, no bedding

Sow in farrowing crate. Note all-slatted floor and rubber mat over lying area

half-inch skim of cement and sand over the top of that; the latter may be omitted if the concrete can be tapped up enough to get a smooth finish with a steel float. I find it just as easy to leave the concrete rough and skim over it with cement and sand the next day; this is particularly so if you have help and if the concrete is coming in thick and fast from a concrete mixer. However, a reasonable concrete floor can be obtained with a load of what we call 'three-quarters-to-dust' chippings, the dust side of it being substituted for the sand; a ratio of four parts 'three-quarters-to-dust' to one part cement, with sand also being used, a ratio of 4–1–1, will suffice.

There are many insulation materials on the market, mainly produced from oil byproducts, which are all suitable for floor insulation. It must always be remembered not to *bury* your insulation too deep as the very effect that you are trying to achieve will be lost; so not too thick a layer of concrete on the top.

The floor will not need much 'fall'. Too great a slope will mean that bedding will be forever coming out of the door. I would suggest a ratio of 1in in 8ft, or just enough so that if the pigs wee in their bed it will run the right way; the same for the dunging area; enough for the liquid to run away, but not enough for the solids to work down and block the drains. The pigs will also appreciate a gentle fall as they will not slip on it so easily.

5
Fencing

Good strong fences are needed to keep pigs under control. In general, the smaller the area they have to run in, the better the fence has to be, as it soon may become a question of the grass being greener on the other side; the bigger the area the pigs have to roam the more contented they will be.

Hedges or Devon bank structures cannot be relied upon to keep the pigs in. Pig netting, barbed wire, galvanised iron or an electric fence must be put up inside the hedge.

Pig netting is expensive but is permanent and durable; it will need a post of iron or wood about every two metres, and if the pigs are not ringed (see Chapter 20), I would suggest a strand of barbed wire along the top and bottom, with a steel peg into the earth in between the posts to keep the wire down. Barbed wire will keep them in, but you will need four or five strands some six inches apart.

Galvanised iron is sometimes used with strands of barbed wire four to six inches above the top of the iron. A four-bar wooden fence is marvellous, but if you find the pigs are chewing the wood, pig netting or chain-link fencing can be nailed up.

Electric fences are being used more and more for stock. If you have a fence that is not quite good enough, one strand of electric fence will do the job. If you want an electric fence across open ground, I would recommend two strands of wire, one eight or nine inches from the ground, and the other one nine inches above that. Insulators can be bought, but can also be made for next to nothing (see diagram).

Electric fence stakes to take the insulators can be of any bits

FENCING

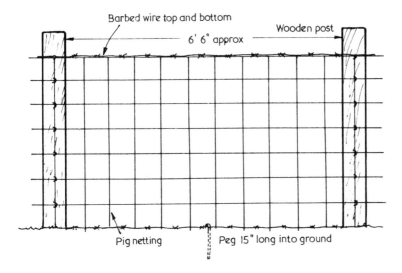

Fencing: barbed wire will keep pigs in but you will need four or five strands of plain wire (6in apart) below it. The pig netting is pegged into earth with 15in pegs. The posts are steel or wood, 6ft apart

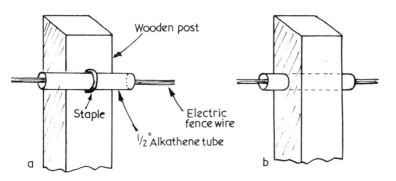

Fencing: electric fence wire run through alkathene tubing, fixed to wooden posts

of wood provided they are 2ft 6in in length and preferably treated with wood preservative. Most people will be able to find wood lying around suitable for this purpose. Cut off lengths of alkathene, just a little wider than the stakes, and secure at the appropriate height with a large staple. Holes can be drilled through the stakes with the right size bit, to ensure a tight fit when the alkathene pipe is hammered through, leaving half an inch jutting out each side of the post.

6
Feeding

There is virtually nothing that the pig will not have a go at eating, from brambles to onions. If left in a patch of ground with every conceivable weed in it, like stinging nettles, couch grass and docks, it will clear the lot, eating everything above ground and rooting up the earth to eat all the roots as well; that is, of course, if your pig has not been ringed (see Chapter 20). You will end up with a completely weed-free patch, already manured and waiting to grow any crop you want. (You will, however, need a good fence around your patch—see previous chapter.)

This is perhaps oversimplified, but it is what pigs are capable of. It is so effortless to them because they are built precisely for that purpose and have been rooting for food for generations in the wild.

If you are endeavouring to get your pigs to clear your ground, when you feed them with nuts, don't put them down on the cleared ground, throw them into the middle of the weeds, stinging nettles and brambles; no matter where they land your sow or sows will sniff them out, casting all aside in the process. A few well-chosen blows with a strong digger from time to time underneath a well-rooted bush will help. Your pig will be reluctant to root in any particular place if there is nothing there for it too eat. Some attention will be necessary to make sure there are no poisonous plants in your plot before you put your pigs onto it (see Chapter 24).

This does not mean of course that you do not have to feed your pigs anything else. They also need a good pig nut or cobs, which are bigger than nuts, used mostly for field feeding. The cobs, being bigger, do not get trodden into the earth so easily, and if you are troubled by birds they are too big for them to carry away. Four to five pounds a day should be provided; once a day is sufficient for dry sows. If you think your sow is too fat try cutting down the nuts a little, and if out on good grazing grass up to one and a half pounds a day can be saved. It is well worth while turning your attention to anything that will cut down your feed bill. By far the largest single pigkeeping expense is food, perhaps as high as eighty or ninety per cent of the total costs.

Something that has always appealed to me is fattening a pig on greenstuffs and waste vegetables from the garden (something for nothing, I suppose), with pig meal as well of course; resulting pork tastes so different, on a par with free-range eggs as opposed to battery eggs. You must be careful here not to violate the 1973 Waste Foods Act, which states that no kitchen scraps may be used which have been in contact with any part of a carcase.

At one time in the country it was quite natural for families to keep a pig in a shed at the bottom of the garden, and a familiar sight hanging up in people's sheds was the pig form used to carry the dead pig, although these were often lent around the village.

It was a big day when someone killed a pig; neighbours would boil water in their coppers to pour over the pig to enable the hairs to be scraped off, and would probably get a joint of pork for their trouble. Some of the meat would be salted down in big earthenware jars, and a leg or two would be hung in the chimney breast to smoke; hog's pudding would be made from the intestines and the blood. I can still see my mother running into the house with a bucket of blood before it got cold and congealed, as it was not good for making hog's pudding if cold. Nothing was wasted—the head would be made into brawn, and this would all have to be done virtually the same day, and with pigs anything up to thirty score (272kg, 600lb) live weight.

FEEDING

The times I have heard the old people say how they would like a bit of fat bacon—that is where it used to come from. A pork pig going to the factory today would be 63kg (140lb) live weight and a bacon pig 91kg (200lb) live weight.

Times have changed and today people who keep their own pigs for eating cart them off to the local abattoir in the back of a van, fetch them the following day, joint the meat up and put it in the freezer.

So we have the pleasure of rearing and eating our own pork, and not having missed too much money spent on the food over the past weeks, and of course we have all the dung to put back on our gardens to grow more vegetables to feed the pigs, and so on.

But back to feeding. When we eat a cabbage from the garden there always seem to be more leaves to throw away than there are to eat; out over the fence they go, into the paddock, where eager sows or gilts are waiting to eat anything we care to throw out to them. So no compost heap for us, which we don't need anyway because we have mountains of the real McCoy.

Feeding your fattening pigs is quite a simple affair, provided of course they have good accommodation and are draught-free and *warm*. No pig, or any animal come to that, is going to 'do' if it is cold.

A fattening nut or meal can be fed on their lying area, provided it is free from dung. They will be quite happy to pick it up from there, water being given separately. A trough can be used to give them meal and water mixed as a swill. On fattening trials or troughs a wet feed system comes out slightly better than a dry one.

Skimmed milk is an excellent feed for pigs, fed in conjunction with meal. Milk is brought from the farmers to the creameries, poured into large trays, allowed to stand, then heated to just under boiling point and left to cool. After this process skimmed milk, as the name suggests, is milk without the fats, the cream having been skimmed off the top. Unless you have a big herd it is not possible to use skimmed milk, as the milk tankers deliver several hundred litres at a time; and although it can be used even when it has 'gone off' there is a limit.

Now if you have a cow, that is a different matter. Milk could then be given skimmed or not, bearing in mind that 'full milk' will be a lot richer and would want to be introduced more carefully.

Whey, a by-product of cheese-making, is another alternative used to ease the meal bill, but with the same sort of problems as the skimmed milk; it would need to be done on a larger scale to warrant the piping system necessary to feed it.

Swill is a very good feed for pigs but needs to be prepared on a larger scale to justify the equipment necessary *by law* to sterilise it. To do the job properly a pipeline would be installed going direct to the fattening pens' feed troughs, far better than lugging swill about in buckets.

There are many pigs fed on swill that is picked up from sources like hospitals, hotels, pubs, large holiday camps and army barracks. This used to be a cheap source of feed, but with the running around to pick the stuff up and the fuel needed to boil it, it is not such an attractive proposition as it used to be. It is not the cleanest of jobs either.

Stringent rules and regulations have now been introduced to swill feeders—there always has been tight control on swill, but since the first outbreak of swine vesicular disease in England in 1972 special licences have been required for the movement of pigs to and from swill establishments. They are always a cause for concern, and with good reason, as meat from practically all over the world finds its way into eating places and hence into the swill bins. If this meat is not sterilised thoroughly it can be a major contributor to outbreaks of swine vesicular, and foot and mouth disease.

It is argued that swill-fed pigs produce a better-tasting pork and bacon, and I am inclined to agree with this; it is probably the wider variety of food in the diet. I am back to the free-range eggs again, where the hen picks up natural vitamins and minerals, such as a maggot or two from the dung heap, a worm, some flies, a few blades of grass, and from scraping in the earth any little insects and grit for the eggshells. So it is with swill for the pigs—natural varied food, something we all need to keep fit and healthy, and different from the artificial minerals and vitamins incorporated in the feed-merchants' food. Some people may scoff at this opinion, but then, why is it that there is this distinct difference in taste between the two?

Pigs are being fed on maize silage, particularly in America and other maize-growing countries, with mixed results. We are growing more maize in this country now, but we can never compete with the major maize-growing countries; we haven't the climate, not enough sun.

Creep Feeding

If you want your piglets to have a good start in life it is important that they start eating creep feed (so called because we start offering in the creep) as early as possible, for the obvious reason that they will grow and 'get away' quicker. Perhaps more important is that the sooner solid food can be got into the piglet's system the better chance it has of withstanding *E. coli* infection. It is a curious thing that on some farms creep will be taken as early as three to four days old, yet other piglets may not take interest for at least a fortnight. All kinds of persuasive niceties are offered to tempt them, and different pigmen have their own ideas on the subject. The pity of it is that it is always the biggest pigs in the litter that start eating first. One would imagine that the smaller pigs, who are not getting as much milk, would be hungrier and start eating the creep first, but unfortunately it does not work that way.

Cornflakes and other breakfast cereals, brown or white sugar, dried milk powder, custard powder—you name it and I will bet it has been tried. All these appetisers are sometimes sprinkled on top of the creep feed pellets.

My own personal procedure on the subject is an idea I picked up during a pig trip to France. It is a mixture of sedge peat or peat moss, milk powder (an excellent food at a relatively low cost) and creep pellets. Only we offer it to the piglets separately, a small handful of each. It is incredible sometimes to see a piglet come and pick up mouthfuls of the peat (usually when it is offered for the first time, at three or four days old), stand there and eat it; they seem to 'go on' all the better for it.

In the initial stages a creep feeder is not much use, a common container for the first offerings is a brick with a frog (hollow) in it, which will give you an idea of the small amount needed to start with.

I like to start them with the creep on a three- or four-inch concrete block, up against the wall beside the light. They seem to take a delight in climbing up onto it to investigate and are quite happy to eat the creep feed there. For some strange reason if you put the food in a creep feeder, they are not interested.

A little food may be wasted to start with, but the amount given each day is very small so it is no great loss. As I have said, the important thing is to get them to eat some solid food as soon as possible, gradually phasing out the peat and milk powder over a couple of weeks. You may find they are only interested in the pellets, in which case just carry on with them.

Water is essential if the piglets are going to eat any amount of food, even though they may only take a sip or two at a time. Indeed, I would go as far as to say that water is more important than the creep feed; a piglet may often have a drink of water, perhaps because it thinks its mother's milk is rather rich.

There is a varied assortment on the market of piglet nipple drinkers which can be fixed next to the food, connected to a header tank.

7
General Routine

It is useful to have a general routine worked out in your mind. You get used to doing the basic thing first, like, feeding, cleaning out and bedding up. These are all everyday essential duties which go a long way to ensuring the welfare and happiness of your pigs.

A good habit to get into is to clean out every pig every day, which makes the job easier and quicker. When a day is missed out there never seems to be twice the amount to clean up the next day, but three times as much, due possibly to more bedding being soiled than necessary.

It is important to keep records (see Chapter 1). When a sow farrows or goes to the boar, write it down immediately so that you know when to wean the litter, or, what is more important, when she went to the boar, so that you will know exactly when she will be farrowing next time.

Check your gates and doors regularly; remember the old saying, *A stitch in time*.

Make sure fences and electric fences are secure and adequate.

Keep your everyday tools up to scratch. There is nothing worse than trying to clean out pig houses with a bristleless broom or a shovel with the handle falling out, or a wheelbarrow with the wheel dropping off. All these things will make for an easier and more peaceful and successful business.

In general it is a case of keep pigs clean and they will have clean habits, keep them dirty and it will encourage bad habits.

But this is not always the case. Sometimes a pig or pigs will make a terrible mess in their houses every day for no obvious reason. There is something not quite right, but it is difficult to put your finger on it. Some of the reasons could be; a cold house, lameness, or coming up to the heat period.

Sows with only a few weeks left to go prior to farrowing, obviously feeling fat and blown out, may not always bother to go outside to relieve themselves.

8
Equipment

For the smallholder there should not be too much expense or capital tied up in machinery and tools.

By far the biggest expense could be a tractor and possibly a dung spreader; these items would never be used solely for the pigs, but would be used for other projects. There is no way (although it may make things easier) that this sort of expense can be justified on a sow or two, but I know what people mean when they say they would rather be independent.

For the everyday cleaning out, a brush (one with nylon bristles will last much longer), a shovel and a wheelbarrow would be the order of the day, plus a hand brush (churn brush) for scrubbing out the water bowls etc. If there are any fairly large areas of concrete to be cleaned up, a squeegy is an excellent tool.

A good pair of snips that would possibly do teeth and tails, some animal markers, a thermometer, a syringe and some needles are other necessary items. A hammer and nails are always useful, and if your pigs will be running out, a slodge (a large hammer), an iron bar (for making holes in the ground) and an axe will come in handy.

9
The Boar

Boars seem to me to have two main aims in life, eating and serving pigs. I sometimes wonder which they like doing best; it is probably a case of eating at mealtimes and serving pigs any other time. Boars are like bulls—they can *never* be trusted. When handling or driving them *always* use a sheet of galvanise, a proper pig board or a sheet of anything; a stick is useless for driving them, and if the mood takes one he can be far quicker than you or me, and a wack with a stick would do no good whatsoever if he decided to turn on you. I can understand a boar's feelings sometimes, especially if he has been waiting all day or all night to serve a pig that might be in the next pen, and knowing that all she wants is his attention; the keepers then have other ideas as to where and when the pig should be served, or might even be contemplating using another boar, which would make matters even worse.

I have in the past had to give a boar a good hiding with a stick or a short piece of polythene tube, because you cannot tolerate a boar that is continually playing up. He has to know who is the boss or you would have to sell him; for it would be far too dangerous to keep him. However, I am happy to say that on the whole boars are very good, and it is very rare to have trouble with them; most are quite friendly and amiable and they love having their backs scratched.

On a commercial scale it is recommended that you have one boar to every twenty-five sows. This is not to say that one could not keep less sows and still keep a boar, but the less sows you have the more expensive the servings become. It is difficult to

THE BOAR

say how many sows would justify keeping a boar, but with just a few sows it certainly would not be economical. In this situation a neighbour's boar might be the answer. The movement of the sow or the boar would need a licence from the local authority, which is easily obtainable. Such regulations were brought in with the first outbreaks of swine vesicular simply to give better information of the movement of pigs in such an outbreak.

Perhaps a better idea than a neighbour's boar would be AI—artificial insemination—something which in recent years has made enormous strides (see Chapter 11).

To serve a sow it is better to take the sow to the boar than the boar to the sow; or better still, what I prefer to do is to take them both out and have them in a neutral pen. It is always best to supervise the servings, making sure that both animals are comfortable, and that the sow has been served. Sometimes a boar will mount a sow and seem to be serving her, but on close inspection you may find he has his penis in the sow's rectum. Most boars know this is not right and will start again. When the boar's penis enters the sow's vagina it goes in past the water bladder entrance and into the muscles of the cervical tube where the corkscrew end of the penis can be felt by the boar to be held tight. It is then and only then that the boar will ejaculate; he will not find this condition in the sow's back passage. Under normal circumstances, if the sow and boar can be matched reasonably well they will mate with no problems. A young boar may not be big enough for a big sow or a big boar may be too heavy for a gilt; this is where a serving crate comes into its own.

The crate can be a simple affair with two fixed sides, a fixed front, a removable board at the back and sloping boards on each side for the boar to rest his front feet on, thus taking the weight off the sow. Remove the back board, drive in the sow or gilt,' replace the board and allow the boar to mount her from behind. If he is young and you have a large sow to serve, before you allow him to mount put a three- or four-inch stand or concrete blocks just behind the back board where his hind legs will be, to give him extra height.

It is difficult to make hard and fast rules as to how often a boar should be used; there are conflicting opinions on the subject. At my local AI centre, semen is taken once a week; we certainly at times use our boars more frequently than that with good results.

Last year we had ten gilts out in the field running with a young boar who I swear served them all within a week. He seemed to be riding them all day long. Well, they certainly all farrowed within a week with good-size litters.

Feeding

It is difficult to specify exactly the amount to feed a boar. If a boar is being raised it is good to have him growing quite fast up to about bacon weight, or when he is reaching maturity and will soon be able to start work. From then on his food must be restricted; there is no question of giving him all he will eat as this would cost a small fortune, and he would be far too fat to work properly. Another reason against overfeeding is that he would get too big too quickly, thus not being suitable for any gilts that may be coming along. So all you can do is to feed him whatever you can get away with, something like four or five pounds a day—enough to keep him in condition, but not enough to make him grow too fast. He can run out on pasture with the sows or gilts, but will have to be ringed like them if you do not want him to root up your field. He can have the same food as them, but you may find he will bully the others at feeding time, so spread the food out as much as possible.

The Young Boar

Try to give a young boar a gilt or a small sow that is well and truly on heat for his first service. A sow or gilt well and truly on heat will just freeze and stand rigid when she gets the smell and sight of the boar; so he will have a better chance of getting the hang of things. *Never* put a young boar in with some old sows, as they will almost certainly go for him if he is new. The boar must *always* be the boss over the sows; if he gets a good hiding when he is just starting to serve pigs it may put him off for the rest of his life, making him reluctant to serve or go near bigger, older sows.

10
Breeding

Like all animals when they come 'in season' or 'on heat' or 'boaring' (some of the terms used), the sow's behaviour differs from normal. She becomes more alert, perhaps talking more, though sometimes her appetite drops off. She may dung and soil her lying area, or bite the door and the rails; all she is really saying is that she wants the attention of the boar; she wants to be mated. Sows riding each other is a sure indication, but it is the pig that is standing, not necessarily the one that is riding, that is on heat.

This reminds me of the joke about the sow that wanted to go to the boar. The farmer put her in a wheelbarrow and wheeled her around to a neighbour's farm which had a boar, served the pig, brought her back and put her in her pen. Next morning the farmer couldn't find her anywhere, till he went round to the back of the barn—and there in the wheelbarrow was the sow!

I know an old lady who keeps two saddleback sows and to my knowledge she has never taken either of them to be served by a boar, because they always take themselves, getting out of the orchard they are kept in and walking the one and a half miles to the neighbour's farm where there is a boar. This is right in the country, you understand, with barely a road between the two places. The old lady will ring up when she discovers a sow missing and say, 'Hello, Farmer So-and-so, have you got my sow over there?' And the farmer will say, 'Yes, she was here this morning when we got up, my dear, and we put her in with the boar and she's been. She won't hurt there until tomorrow, you can come and collect her then.'

BREEDING

There will never be any money changed hands; payment will be a few dozen eggs, an oven-ready chicken or duck, some choice apples, or whatever fare she has that the other farmer needs.

This demonstrates to me the enormous smelling ability or instinct of the pig, something they obviously needed in the wild. However much pigs are bred, crossbred and interbred in man's constant endeavour to breed the pig that we are supposed to want, the natural instinct of the pig will take thousands of years to breed out. I will give you another example by referring back to the battery hen. (I do this because I see the pig following the chicken's progress exactly, in the name of economics, efficiency and productivity, going from free-range to deep litters and then to battery cages.) A couple went to a battery-hen farmer and wanted to buy half a dozen of the hens, so that they could let them enjoy their life as they were meant too. The farmer laughed at them. 'What!' he said. 'They would be lost, they wouldn't know what to do with themselves, having been in cages all their life.' Nevertheless he sold them some, which they duly took home and let out; within half an hour the hens were all dusting in the earth.

Pigs are already being put into wire cages with conveyer belts underneath to take away the droppings. At the time of writing this book, there has actually been a law passed in West Germany, brought about by pressure from animal welfare societies etc, making it illegal to keep hens in cages, although there is some confusion as to exactly what this means. If this trend continues and spreads to other countries, as I am sure it will eventually, it will have a major impact on the production of eggs in this country—far dearer eggs, for a start—and will mean millions of pounds' worth of equipment being made obsolete, and what is more relevant to this book, any legislation on this subject is bound to include pig production. It could well put a stop to modern ideas like keeping pigs in wire cages. It is a fact that keeping animals in these conditions is not natural.

In the reading of this book you may have the impression that

I am against modern methods of keeping animals. The difficulty arises when one hen, for example, decides to pick on another; after drawing blood, the others in the cage join in and the poor thing is literally eaten alive—the hen has to stay there and take it, as there is nowhere to go. If they were running free such situations would not occur; this to me is where the cruelty comes in. So with ever increasing stocking rates, and particularly with pigs in wire cages, the same situation is going to arise. We used to keep thousands of battery hens; in the mornings when we picked up the eggs, we could not hold a conversation without shouting because of the din created by the hens singing at the top of their voices. There could be no doubt in anyone's mind, if they had had any experience with animals, that these hens were not happy.

I have been in dozens of weaner houses with a controlled temperature of up to 80°F, controlled air flow, and with subdued lighting—which is known as a 'controlled environment house', where pigs are weaned mostly at three weeks old and put into these weaner houses on expanded metal or concrete slats, or on wire with no bedding at all, with wire or board divisions, which enable the stockman to observe them quite easily. The inescapable fact is that these pigs are happy and quite pleased with life; except for sleeping and eating they seem to be running, playing and dancing all the time, provided they are fit and well, of course. Now with the high cost of heating, this type of housing (three or four tiers of cages) is being used so that people can keep three or four times as many animals with the same heating costs.

Sows sometimes get very thin, what is known as 'pulled down' or 'fleshed off' by the litter; so when they have been weaned, it is sometimes wise to forgo the first heat period to give them a month or so to put on some weight before they go to the boar. The cost of keeping a sow barren for an extra three weeks would amount to little more than the profit from one piglet next time she farrows.

When still with the litter, sows will sometimes come on heat.

BREEDING

It is not clear why this occurs, but it is more frequent a few days after a sow and litter has been moved (ie into a multisuckling house). There is no reason, if in a fair condition, why the sow should not be mated at this time. The sow is more likely to 'return' than if mated after weaning; but there is nothing lost if she does not conceive, as the heat period would not occur again until three weeks after. From my records sows that do conceive whilst still suckling have good litters, both in weight and numbers.

Breeding your own gilts is both pleasurable and interesting, and can give immense satisfaction. First you must select the sow from which you hope to breed your gilts; some of the attributes you would be looking for are as follows: a sow that has a proven record for throwing good litters, and is a good mother and 'does' them well; a sow with good length, good hams and no trouble with her legs (which should not be too long), fine shoulders, small head and ears. I am not saying that you will find all these characteristics, but that is the sort of pig most acceptable today.

By and large, gilts can be selected at a few weeks old; this is my practice, so I can tattoo a number in their ears. Selection of the dam and boar would already have been made with the intention of keeping the offspring for gilts or boars. The characteristics they have then will still be evident when adults. A most important factor in the selection is the number of teats which are clearly evident. No gilt should be retained for breeding unless she has at least fourteen evenly spaced teats. This is in fact the rule rather than the exception. On the whole I would say boars are more highly selected than gilts; so the chances are, whether he is yours or anyone else's, a boar will be of good stock. If you are contemplating an AI boar you will have a selection, and some details are given with each one. If your chosen sow is a little short of length, ask for a boar with good length, and the same with hams or whatever.

Gilts can be reared with other fattening pigs and require no special treatment, but if at all possible, do try to get them out on grass about a month before they go to the boar. This hardens them up and helps to set their bones correctly, and generally makes them fitter all round for the life that you have chosen for them in the years to come.

Gilts can be served when quite small, and if fed properly will have nearly four months to grow on after they have been served. The only drawback to serving gilts too young is that after they have farrowed they are just not old enough to have built up sufficient resistance to the *E. coli* bugs. Consequently the *E. coli* affects the litter through the mother's milk. To counteract this, it is a good idea to take some droppings and cleanings after the older sows have farrowed and put these into the gilts' pens, before and after they have been to the boar. In this way their systems have at least been exposed to the bugs and can build up some immunity against them. On the other hand, gilts that are too big when served for the first time tend to get too big too quickly, the disadvantage being that they will need more food to sustain them and are more likely to overlay and suffocate the piglets at farrowing time. Gilts are not generally served at their first oestrus; if grown well, they will not show any heat periods much before they reach 90kg (200lb). It is not advisable to serve them before they are two hundred and twenty pounds or more, or getting on for six months old. If running with the boar at this time, he will serve them when they are on heat; try to keep an eye on them so it can be recorded when they are served. Note the date that the boar was first put in with them, if you have not seen any activity from him. As I have said before, it is important to know when they are to farrow.

Knowing when a gilt will take the boar is easy to me, but some people find it very difficult. She may not show any of the normal behaviour of a sow, as this is a new experience for her. The surest indication to me is the swelling up of the vulva to a deep pink colour, which happens over a two- to three-day period. Although she may show interest at this time, it is not usual for her to take the boar until one or two days *after* the

BREEDING

Breeding: pressing the hands down the sides of a sow's back to find out if she is on heat and will stand to the boar

swelling has started to go down. A boar will take all the worry and frustration out of the problem. I can appreciate the problem of having to decide categorically that a gilt is on heat, especially if she has to be taken away to a boar, or a boar has to be brought in or AI ordered. A prick-eared type on heat will, when touched, raise her ears in anticipation. One can try to see if the gilt will stand by pressing your hands down one each side of her back (see diagram), but I prefer to sit astride her and dig my knees in, trying to be as rough and as heavy as a boar. If she will stand to this treatment it is obvious that she will stand to the boar.

The most difficult pigs to get into season are second-time gilts, which are gilts that have had one litter and have been weaned. It is not known why they prove to be so reluctant. It is one of those mysteries that vary from farm to farm. Everything is done to induce them, like having a boar next door or in the same pen. Mixing them with older sows has done the trick for some people (not permanently, just for the day); if you are unsuccessful consult your vet, who can induce her with an injection.

The question of how long to keep your sows, and how many litters to let them have, is a debatable one. In the main commercial herds, sows are kept for two and a half years or six litters, and then the trend is for culling them younger rather than older, perhaps influenced by the continuing high prices paid for sows. The price paid is more than enough to replace them with gilts if home-reared. This is not to say that the sow kept in the paddock should follow suit; indeed, she could well have ten litters and more if she is enjoying good health, and health is the decisive factor.

The smallholder keeps sows longer on average than the commercial herds.

Commercial pig people take the view that every litter after six is taking an unnecessary risk, and a long hard look is taken before the sows are kept on; often a programme is drawn up whereby they are culled automatically after X number of litters. Recently I was reading of someone who was rearing gilts, letting them have one litter and then culling them, but I fail to see the logic of this. The inescapable fact, as most of us know, is that there is no substitute for youth. All I can say is that it is a gamble to keep sows too long—from the point of view of a greater risk of complications in farrowing and poor or small litters.

11
Artificial Insemination

AI has made tremendous strides in the last twenty years or so. I remember when it first started; out of eight pigs I had inseminated only one was successful, and she had a small litter, at that.

Today results from AI are just as good as using your own boar, both in conception and in numbers born. This is an enormous step forward in genetic improvement, and progress for the health and wealth of the pig industry. I have particular cause to be enthusiastic about AI, as over the years all the illness and disease that my pigs have been unfortunate enough to have contracted—virus pneumonia, rhinitis, etc—has been as a direct result of buying in gilts and boars often from reputable establishments.

So now on my farm we have vowed that we have bought our last pig, and are rearing all our replacements from artificial insemination.

It tickles me pink to think I can pick up the telephone and order, and a few hours later can collect from the nearest railway station semen from some of the best boars in the country.

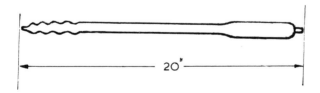

Artificial insemination: 20in catheter. It is slightly flexible

ARTIFICIAL INSEMINATION

It is not hard or complicated to administer AI, and with a little tuition from the AI specialists the process can easily be mastered; there is no reason why you should not be successful at it (see diagram).

Frozen boar semen, which a few years ago was thought to be impossible, is now being imported from the other side of the world. This new technical development will, I am sure, have a major impact on the pig industry.

12

Farrowing

This is without doubt the most important part of pigkeeping. If you do not get the farrowing right then all your efforts will be in vain. If your sow does not rear a good number of piglets, the food she has eaten during her four months' pregnancy will be very expensive for you, plus your time looking after her and perhaps money layed out in the housing of her. Pigmen today would be looking for at least ten piglets a litter, and a little over two litters a year per sow. Gilts are not expected to have quite so many, and it is generally accepted that eight piglets is a good-size litter for them to start off with; but they can and do have more, and can rear them perfectly well. Anything below these figures would not be considered good or economical.

Bring your sow into her farrowing quarters about a week before she is due to farrow. This will be a precaution in case she farrows early, and she will be able to get used to her new surroundings. It is also necessary to worm her a week or so before she gives birth (see the section on worming in Chapter 23). If possible give her the worming mixture a couple of days before you bring her in, so that any worms that she may have will come out in her dung in the field and will not be allowed to infest your buildings or farrowing area.

If you have a spare pen or a yard where she can be kept still enough, give the sow a good scrub with warm water and soap or detergent before you bring her in. Most pigs love to be brushed and scrubbed and will oblige by lying down. Give her udder and all around her backside particular attention, as this is the area where the worm eggs can be laid ready for when the piglets start to suck, and so pass the worms onto them. A good mange wash can then be poured all over her with a watering can to make sure your sow is free of mange, lice, and worm eggs. Not all pig farmers practise this procedure, but if you want to do the job properly and leave nothing to chance, I suggest you scrub every pig you bring in to farrow. If it is not possible to give the sow this treatment away from where she is to farrow, then do it there and brush it out well afterwards.

People do of course farrow pigs in all types of housing; some just use a square shed with some sort of farrowing rails around the sides, which are most important. Others have solari-type housing with the sloping roof down to the back where the creep would be, perhaps a heater light, and again the farrowing rails.

Too much straw is not to be recommended. No matter where you put it the sow will have other ideas and will rake it all up in one heap around her; when the piglet is born, especially if it is cold, it will go under the straw to keep warm—the sow will get up and then lie down again and, Bob's your uncle, one dead piglet!

This reminds me of years ago when I first started farrowing

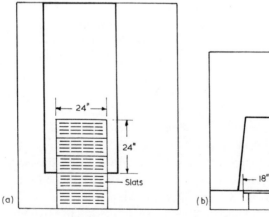
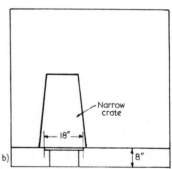

Farrowing crate: (*left*) seen from above and (*right*) end-on. See also photograph and following diagram

FARROWING

pigs, and in this type of housing. This particular house had a four-inch step down to the dunging area, with disastrous results. For, one very cold morning, I discovered the sow had farrowed and there were four or five dead good-size piglets down in the dunging area; they either could not get back, or were too cold to so, a bit of both I would have thought. It was enough to make you weep. They say you learn by your mistakes; I do not believe there is a truer saying.

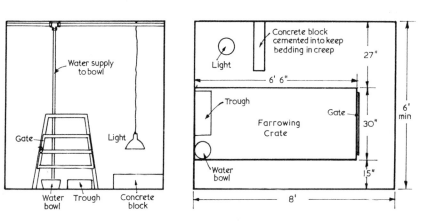

Farrowing area: (*left*) seen from end and (*right*) layout

For me there is only one way to farrow a pig and that is in a crate, with an infrared lamp by the side; it is then so very much easier to look after her. Injections can be given comfortably without trying to keep up with a sow running around a pen, and any other little problems can be dealt with— like a bit of canker in the ears, a sore place, or perhaps her feet need a bit of attention. All these things can be done in a farrowing crate, and it is useful to get your sow in a crate for ringing (see Chapter 20).

Farrowing crates are now used by nearly all commercial pigmen, being accepted universally as the best way to farrow pigs. The beauty of a farrowing crate is that it can be set down in a corner of a shed, or can be situated in a big shed, with a portable roof over the top for warmth if necessary. The sow will be in the crate and the piglets will only require a light boarding around the outside, no higher than eighteen inches.

Three or four days after farrowing, mother and babies can be moved out and the crate removed if necessary, particularly if the crate is being used for one sow. Several sows would need more permanent farrowing crates. A second-hand crate can easily be obtained and it is certainly something that does not lose its value.

More piglets are lost by being lain on and suffocated than by perhaps all the other fatalities (at this age) put together.

Within twelve to twenty-four hours of being born, the little piglets will get to know where the heat from the infrared lamp is, and after each feed will go straight back and lie down under it, thus preventing the sow from lying on them when she gets up and down. If you are on hand after the piglets have had their first feed and you see them just snuggling up to the sow's udder, pick them up and put them under the light, which will get them to know even quicker where the warmth is. The only time that I have found the piglets not interested in lying under the light is in the summer on very hot days, when they are quite content just to lie around anywhere on the bedding; then it is a waste of money having the light on, but during an English summer this does not happen often.

If you can keep your litter alive for the first twelve to twenty-four hours you are over the critical period.

It is very important to know when your sow is going to farrow, so I will mention some of the antics that she will get up to, normally anything up to twelve hours prior to giving birth. Sometimes you can walk into the farrowing house and there is the sow really working at it, walking backwards and forwards as much as she is able in her crate, scraping the bedding backwards with her feet, pushing it forward with her nose, picking up the straw in her mouth and chewing it up, putting it here, putting it there, biting the bars and generally being very busy

FARROWING

instead of being her own peaceful self, just lying down comfortably, except at meals times of course.

This is to a greater or lesser degree how most sows react when farrowing is imminent. There are other little signs that may indicate her intentions, like playing with the water (if she is not prone to doing this) and wasting it. Most pigs drink more when they are due to farrow, perhaps because of labour pains they feel the need to drink more. If farrowing is imminent a sow may not eat all of her last meal — but only the very last as a rule, unless something is wrong. You may also find her standing up with all her bedding back behind her, or she may just be standing up more than usual.

Try milking her teats. If she has an abundance of milk, that is a sure sign that she will farrow within twelve hours. Some older sows, however, may have a drop or two several days before, but this is no indication. If the pregnancy is normal, pigs are good to their farrowing time and are not usually a day or more out either side of their dates. When you are sure that farrowing is imminent do not under any circumstances give your sow food, even if it is her feeding time, or however much she is asking for it; there is no worse sight than seeing a sow being sick in the middle of labour, brought on by all the heaving and pushing.

The first signs of the immediate first-born will be a small amount of transparent mucus coming from the vulva. Many a time I have been into the farrowing house, perhaps last thing at night, not having checked the sow for several hours, and have just spotted this telltale sign, enabling me to put the heat light on and give her extra bedding, especially around the backside where the piglets will be scrambling around, to stop them getting cold. This you can only do if the sow is in the crate.

A point to remember here is that when it is obvious that the sow is going to farrow, put all the straw around her and put the light on. Do not go in there a couple of hours later when she has got half a dozen piglets and say, 'Oh, she's started,' and proceed to give her more bedding; the least rustle of straw at this stage is all she needs to get up and start pushing her bed around again, for in the middle of labour she won't worry much whether she lies on a pig or not.

I would suggest ninety-five per cent of pigs will give plain, visible warning when they are about to have their litter. The other five per cent will make no move at all to give any indication of their intentions and will simply lie down and farrow.

The actual length of farrowing time varies, of course; how like humans pigs are and, I suppose, all animals in this respect. Some will have them quick and effortlessly like shelling peas, and others seem to toil and struggle for hours on end. Some sows will have a dozen piglets in less than an hour, all finished; the average will be perhaps two to three hours, but sometimes they take up to twelve hours. If there are complications farrowing time may be twenty-four hours or more, which is frustrating and worrying for man and beast. In this case it will be advisable to call in the vet. These occasions should be few and far between, I am glad to say.

When a pig is going to give birth within seconds she will waggle her tail. It is the practice of some pigmen to dock (cut off) the sows' tails when they are born, so the tail waving will not be so evident. I used to tell my children that the sow was giving the piglets the 'all clear' and waving them through.

Some sows will lie quite still during farrowing, but others will tuck in their hind legs and generally roll about. This is something I do not like to see, as injury to the piglets that have already been born may occur.

If you have a situation where the sow is up and down, having a piglet and then getting up, it is far better to put the little ones in a box beside the light to prevent them being stepped or lain on. I have had a sow like this from time to time; if she is very bad and you cannot for one reason or another stay with her, the piglets will come to no harm whatever if left in the box for six to eight hours, or even longer.

Sow houses at Stoneleigh: six sows in each with covered lying area. Gates enable tractor-scraper to clean down the dunging area. Individual feeders

Sow stalls: no bedding, slatted floor at rear

Fattening house. Flaps can be closed in cold weather

Farrowing: breach birth

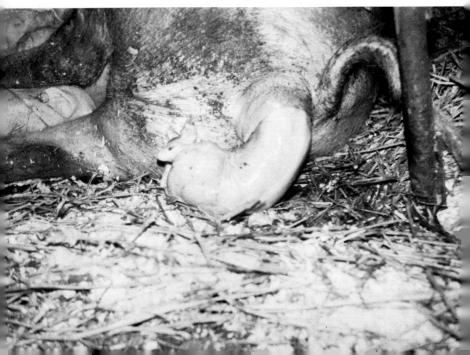

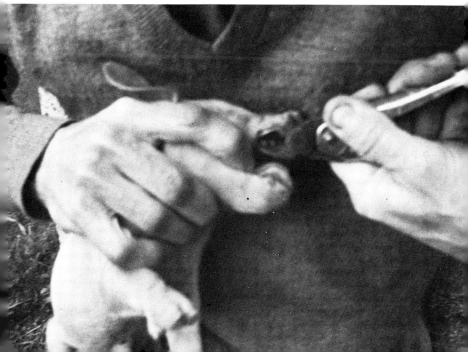

(*left*) Correct way to hold a small pig; (*below*) Teeth cutting: note the two teeth between the snips

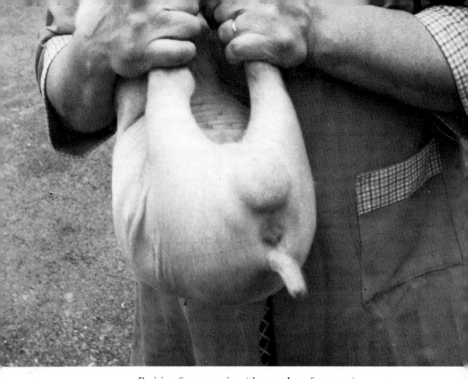

Position for castration (the one I prefer to use)

Another position for castration

Pig with atrophic rhinitis. Note the twisted nose

Fatteners: Landrace crossed with Gloucester Old Spot

Farrowing crate with slatted rear end. Note the concrete blocks on left, cemented in, to hold the bedding, and at the side of the crate, to start creep feeding

Sow and litter in farrowing crate. The bottom bar is adjustable – useful for gilts

Hardy young Tamworths

Large Whites crossed with Landrace maiden gilts

Contented gilts in the paddock

FARROWING

In the process of farrowing, some sows, trying I suppose to make their teats available to the little ones, will lift their udder up several inches—thus allowing the piglets, in their eagerness to find a teat, to push in too far underneath; when such a sow lets her udder down again, the piglets suffocate, trapped underneath her. You cannot be present all of the time, however, and all these risks are in the lap of the gods. It is always a wonder to me that more little pigs do not get stepped or lain on. All piglets have the built-in sense to jump clear as quick as lightning if anything touches their backs, even when only a few hours old.

Some of your piglets will have been born backwards (breach birth) but this seems to make no difference to the mother or baby. The afterbirth (the bag the pigs are wrapped up in) should come out after all the litter has been born, but sometimes some will start to come out in the middle. You will do no harm if you pull it away to save the unborn pigs from getting caught up in it when they are born. From time to time a piglet will come out still in its bag and will kick and struggle to get free, or suffocate. It is sad to find a piglet dead in this condition; you do not know whether it was born dead or just could not kick hard enough to get free. I make no apologies for going into such detail regarding farrowing. As I have said before, this is a very important part of pigkeeping; and as the economists are always telling us we must try to produce more piglets per sow per year, to offset those ever-increasing costs.

A piglet's umbilical cord does not usually come clear when it is born and it as well to pull it out when the piglet starts to move off, so as not to damage the navel. Some farmers make a point of cutting the cord, leaving a few inches, then tying it with string, but I have never found that they come to any harm by leaving it, and in a few hours it shrivels up like an old leather bootlace.

The question of whether to supervise every farrowing is a debatable one. Certainly if your sow is farrowing in the daytime it would be advisable to be there to make sure things are proceeding satisfactorily. However, at one or two o'clock in the morning, one is apt to persuade oneself that she really will be all right if you go to bed, especially if the farrowings have been going on for several nights running; but I do know people who always sit with their sows until they have finished farrowing, whether it be night or day.

About five minutes after a piglet has been born it will be dry enough to hold to cut its teeth (see Chapter 17). Again, this is a preference, but to me it is essential, as the sow lies so much quieter after this is done. After all, the aim is not to disturb her, not to have her jumping up every five minutes because the little ones are biting her teats; and the little devils will. You sometimes pick up a piglet that is squealing and find that another will come up with it with its teeth firmly embedded in its lips. You can hardly blame them, I suppose, as that is the way of it in the animal world—the instinct is to fight for food; the survival of the fittest.

The little ones will in a very short time sort themselves out as to which pig has which teat, and it is generally found that the biggest, strongest pigs will have the teats nearest the head of the sow because these teats will have the most milk. The unfortunate small pigs will have to be content with the ones at the back. The piglets will keep the same teats right through the suckling period. There is quite often a small pig in a litter, variously called a nestlebird, a nestledraft, or a runt; from the outset it is at a disadvantage because it is not strong enough or big enough to compete with the others.

A very good friend of mine, an old-age pensioner who looks after my pigs when I am lucky enough to have a holiday and who looked after pigs all his life on the farm where he worked, would say of a runt, 'They ban worth a damn, boy.' I suppose up to a point he is right; runts will probably cost you more in meal to get a saleable size than they are worth.

So you now have a good litter, I hope, going back under the light after each feed. Now is the time to watch out for *E. coli* in your little pigs (see Chapter 23).

FARROWING

Sometimes less than twelve hours after the piglets are born they will get scour diarrhoea—not just runny yellow diarrhoea but so bad that it is just like water, and all that can be seen is a wet behind. A sign of this is yellow diarrhoea on the backs of some of them caused by rubbing up against each other. It is not good enough just to stand back and say, 'Oh, they look all right.' You must get in there with them and pick each one up in turn, lift up its tail and inspect it, marking it with an animal marker so that you know which ones you have looked at.

I would suggest twelve to twenty-four hours would be the best time to cut their tails. Tail docking is a simple procedure, best done with a good pair of secateurs. I use the same good-quality snips that I cut their teeth with; they should be sterilised before use. There is no need to cut the tail right at the base, about half to three-quarters of an inch can be left. It is quite a simple operation: just catch hold of the end of the tail, decide where you are going to cut, and snip, it is as easy as that.

Big fattening units are reluctant to take weaners that have not been tail docked; the fact is that when weaners go into fattening pens, often with a slatted floor, or into solid lying areas and slatted dunging areas with no straw, they get so bored that they look around for something to do and try biting the tail of the pig nearest them. At one time we would receive an extra twenty pence per pig for tail docking, but nowadays if they are not done twenty pence will be deducted. It is the end of the tail that fatteners go for and of course it bleeds, which does not do the pigs any good; the important factor is that they smell and taste the blood, which is likely to cause them to fight—and once they start that, one or more pigs can be lost in a very short time, which no one can afford. If your circumstances are such that you will most likely sell your weaners it will be a good idea to dock their tails. If you are going to fatten them yourself in the traditional way and give them straw bedding and greenstuffs, there is no reason why they should not be free of this habit.

Two or three days after birth or even earlier your litter can

FARROWING

be injected with an iron solution to stop them becoming anaemic. This is a routine procedure with all pig herds kept intensively with no access to the land and the earth. There are a number of iron products and your vet or veterinary supplier can provide you with these, also with the syringes and needles to do the job.

Iron Injecting
If you intend to run your sow and piglets outside with access to the earth after a week or so, I would suggest that the iron injection is not necessary, as they will pick up what iron they need from the earth. Likewise if you are prepared give them a sod of grass every other day. For a couple of weeks this will be quite sufficient iron for them, and greenstuff is rich in iron.

To give the injection, hold the piglet's right leg in your left hand (if you are right handed). When you inject the pig and start pushing the iron in, try to keep the fingers of your right hand with the syringe in it pressed up against your left wrist so that your two arms are as one; it is sometimes difficult to keep the needle in place when the piglets wriggle, as they often do quite vigorously. With the syringe in your right hand, use your left thumb to push the skin in towards the left leg and inject the pig with the skin still taut, the idea being that when you take the needle out and release the skin the iron is trapped inside. It is recommended that you use one needle for injecting and leave another in the refill bottle, to protect what is left in the bottle against contamination.

13
Weaning

At what age to wean your litter is an arguable point, and a case can be made for all weaning ages.

Firstly there is the traditional eight-week weaning, still practised by some, mainly with sows and litters that are running out. The sows do not get fed up with the litter if they are free to roam around. When they are finally weaned and brought in at eight weeks the weaners fatten well, as they are fit and healthy and their bones well set, due to the running-free life they have led previously. Sows readily come into season after the eight-week break, and it is argued that more pigs are conceived in this system even though the farrowing 'index' is not so good.

Five-week weaning is probably the most popular, although an increasing number of breeders are switching to three-week weaning, with improved creep feeds and better housing facilities. The breeders and the pigs can usually cope with the five-week system adequately, weaners 'going on' comfortably without too much of a check, and the sows coming to oestrus within seven days, thus gaining three weeks on the eight-week weaning, and six weeks on the two farrowings in the year—forty-two days in which the pigs have to be fed, housed and looked after whether they are producing piglets or not.

Three-week weaning or earlier is not to be recommended for the beginner and should not be undertaken until a five-week system is working well. The arguments for three-week weaning are that up to two and a half litters per year can be obtained if the system is working well. Arguments against the three-week

WEANING

system are that the sows do not always come on heat as quickly as in the longer suckling systems, and the litter size and numbers may not be as good. So unless you are getting good results, what you gain on the roundabouts you lose on the swings.

Whatever age you wean, meticulous care must be taken in the amount fed to the weaners especially if the pigs are indoors, as they are more likely to get the scours if they have not been used to running out on pasture. Whatever you were feeding them before weaning should be cut down by, say, a quarter when weaned; *do not* increase the food for a week. However hungry they seem to be, do not be tempted to give them more, as the majority of weaner scours are caused by overfeeding, at a time when, having just been weaned, they have not previously had to cope with just the meal or pellets.

Stress is a major factor at this time, so always take the sow away from piglets, never the other way around. Do not castrate, worm or move them, change their diet, mix them with other pigs, or do anything that is out of the ordinary. Attention to all of these factors will go a long way towards settling the weaners down and ensuring trouble-free pigs later on.

14
Modern Pigkeeping

So-called modern pigkeeping has in recent years made enormous advances into a world of steel and wire, tethers and stalls, slats and slurry—all in the name of progress, presumably.

Pigmen today are keeping growing numbers of pigs with the same labour force that would have looked after a fraction of these numbers a few years ago. All improvements are made with a view to making it easier to look after more.

Thousands of pounds are quite happily spent to this end. You see and read of people putting up new pig set-ups which may cost £100,000 or more. In my opinion this sort of enterprise can never be justified on the evidence of the pig returns over the years; what with the high interest rates at present, they will never see their money back. Perhaps I am just old fashioned, but if I had that sort of money I would put it into a building society, and keep a couple of sows in the paddock.

A typical modern piggery might be a sow house for two or three hundred sows, perhaps with no straw, being tethered or in stalls with slats behind and a slurry channel virtually eliminating cleaning out, or with a solid floor cleaned out with a scraper and tractor daily. The sows will be put into the farrowing house to farrow, perhaps in tethered crates, the crate part of it being virtually a farrowing rail down each side of the sows; or they might be in crates, where they will stay for three weeks after they have farrowed. The three-week-old piglets will be put into a weaner house with all slatted floor, artificial light, controlled heat and ventilation, and will stay here until they are around 25kg (50 or 60lb) in weight, then be transferred to the fattening unit which may be all slatted floor or solid lying area. The sows will go into yards for four to seven days, during which time hopefully they will come in season, be served and then go back into the sow house again. This sort of system needs expert pigmen who know their job inside out, and know what to look for to avoid any trouble. All too often in these bigger units the sheer quantity of pigs being kept in a confined area can lead to disease problems; after all, the pigs are being kept in an unnatural environment which can very soon lead to disease outbreaks becoming epidemics. Then the only solution is to clear the lot out and start again.

Situations like these are the exception rather than the rule, and there are some excellent units up and down the country which keep on top of the problems. The standard of stockmanship is high, and needs to be in these intensive units. One thing that pleases me more than anything in writing this book is the fact that those of you who are going to keep the odd sow or two should never come across any of the diseases encountered by the big intensive people, unless of course they are brought in by people, boars, or other pigs and animals. Intensive pigkeeping is different: wherever there are large numbers of animals or people living in a confined area in poor conditions, there you have *trouble*.

15
Bedding

There is a wide choice of bedding or litter for pigs these days. Like everything else it gets more and more expensive, which is one of the reasons why some pig farmers have been prompted to do away with bedding altogether. This means that all the pigs will be on concrete, steel or plastic slats, or possibly lying on rubber mats, wood or concrete. This is a modern slurry system, with wide channels and all the pens leading to the slurry pit, tank or lagoon, which must be emptied with a slurry tanker periodically, come rain or shine.

Bracken (Fern)
This has been used for thousands of years for bedding purposes, and is still used by those who live near or have access to it. Bracken is cut in summer, allowed to dry and then bailed up; or it is gathered up in the winter through to early spring, before the new fronds begin to shoot. Bracken is a perfectly acceptable bedding and will rot down on the dung heap; there is no food value in it and the pigs will not eat it.

Peat
Peat is a very good bedding material, used extensively in the past, but has now become very expensive for this purpose. It is a marvellous material for putting on the garden or fields with the dung added. Peat can be obtained in loose bulk loads. You may well be fortunate enough to live in the vicinity of some peat works.

BEDDING

Sawdust

An excellent bedding material with good absorption qualities, sawdust is commonly used for bedding pigs, especially for farrowings; it is no longer cheap. You may be able to secure some from a small sawmill for next to nothing, if they are not geared up to sell like the large places are.

Shavings

Similar to sawdust, shavings are three times as absorbent as straw. Some would argue that shavings are no dearer to buy than straw. I prefer shavings for piglets in the creep as they stay put better than straw, which seems to get pushed out. In the past, when straw was in short supply, I have bedded sows on it and on occasions they have eaten every scrap. I was a bit concerned at the time but they all farrowed well with no apparent ill effects. It has been said that if too much is eaten the piglets when born may loose blood from the navel cord, due to the resin in the shavings thinning the blood.

Wood shavings usually come in bales, or wrapped up in polythene. Here again you might be able to get a supply from a small local woodworks.

Shredded Paper

A brand new idea this, with the invention of a machine that will shred paper, for the most part newspaper. This encouraged some enterprising people to start up a business in doing just that—collecting old newspaper, shredding it and then baling it up and selling it as bedding for all animals. It makes a very soft bed, goes a long way, and absorbs moisture very well. There was some concern at the outset that the print on the paper would be dangerous to the animals, but this has not been proved to be so. The only argument against it is the price, but probably no dearer than shavings or straw.

Straw

Straw is by far the most common form of bedding, and it has

one important advantage over its rivals: the pig can eat it. Barley straw is probably the most palatable, being of a softer nature. It is used far more for bedding than either oat or wheat, simply because more barley is grown in this country. I do not recommend anyone to wear a woolly jumper whilst handling barley straw, because of what my wife would call those 'damn barley isles'. Looked at closely, it can be seen that the actual isles have tiny hooks all going one way, so that when they get caught on your clothes, every movement made pushes them further in; they are soon pricking irritatingly into your skin.

I consider straw to be an important part of the health and well-being of my pigs. I may be old fashioned, but I do not relish the thought of keeping pigs without it. I can just see them looking at me and saying, 'Come on, then, where's my straw?' Straw used as bedding makes the pigs more content, relieves boredom by giving them something to do all day in chewing it up (thus reducing the risk of fighting), plays an important part in the diet of the pigs in the form of roughage, and also helps in keeping the skin clean. Pigs with no bedding often have a skin problem, especially sows in stalls or on tethers, who cannot rub up against each other.

I know some eminent people in the pig world who would not agree with these points, and say that straw eating is a habit and not necessary; but as far as I am concerned, as long as our pigs are healthy and we continue to get good results I will not begrudge them the straw. I am quite happy to give a group of fifteen to twenty sows a bale a day. A good half of it will be eaten, a small proportion will get soiled and go out in the dung, the rest will stay as bedding. As a rough guide a bale of straw (if bought on the field at a cheaper rate) is equivalent in cash terms to $2\frac{1}{2}$kg (5lb) of pig nuts. Appreciably less will be eaten in summer, especially in very hot periods.

Oat straw is of very good feed value for pigs, and also for bullocks. Most years we get oat straw off the field. I often think that it is too good for the pigs to lie on, with its lovely golden colour, but my wife says, 'Lovely stuff, no barley isles.' Oat straw can be a bit matty, due to being long, but the pigs love it.

Wheat straw is not considered as good feed value as oat; on the whole it is more brittle, but is widely used for pigs, possibly because after it has been combined and baled it becomes very bitty with more husk—what in days gone by we called more 'dowse', what a lovely word that is—ideal litter for pigs. Wheat straw generally sells for a few pounds less a ton than barley and oats.

Pigs are like us, they have their favourite foods. As far as straw goes, if you could ask them which straw they preferred, I am sure they would say, 'Barley one day, oats the next and wheat the day after that.'

16
Dung – How Much?

It is impossible to say how much dung a specific number of pigs will make. It all depends on how much food they eat and how much bedding is thrown out with the dung. To give a few examples: a group of about fifteen sows will produce a large wheelbarrow load a day with not much soiled bedding in it; eight or nine sows and litters will also produce a large barrowful; so one sow and litter would produce something like a wheelbarrow load a week.

Six fattening pigs could mean two barrow loads a week, but it really depends on how much bedding is mixed with it. The pigs that used to be fattened at the bottom of the garden would not be cleaned out from the time they went into the house until after they had been slaughtered, though the practice is different today. These pigs were quite happy to lie at one end of the house and dung at the other, with straw being given when necessary. This way of fattening a pig was done more in the winter months, when there were no flies about and the manure would not smell so much; being in a heap, the dung would heat up and keep the pig warmer.

Fresh dung is quite bulky if there is bedding mixed with it. After being put on the dung heap for a day or two the bacteria begin to work, and the dung heats up and reduces to about half the size it was when fresh.

Pig dung, manure or droppings—whatever you prefer to call it—is considered to be of good value, as dung goes, for vegetables, fruit, flowers, grass, etc, falling in between horse and cow dung. Horse dung is of a more open and lighter texture, and cow or bullock dung has a far denser, heavier quality—greasy may be the appropriate word for it.

By and large, people are pleased to get any dung they can for any crops they wish to grow. Rotted dung is obviously the best because bacteria have already worked through it, but I have used fresh dung for many purposes with no ill effects.

In the autumn we use dung straight from the pigs' house, mostly with straw and shavings in it, for putting around all the fruit bushes, strawberries and spring cabbage, and a fair whack at that, several inches thick. It takes a bit of time putting it around, but we feel it is well worth while as it is not only food for the crops but smothers the weeds. Come springtime we are left with a nice strawy cover; and in the case of spring cabbage, when they are ready, we pull them up whole—root and all—and are left with nice clean ready-manured ground, to rotovate and sow again.

The one thing that puts me off sow tethers, stalls and farrowing crates is that some sows will push their noses forward as far as possible to dung and urinate, and tuck themselves up, thus dropping it right where they have to lie and so always being in a filthy state. This results in soiling, using more bedding, and a greater risk to the health of the sow, and takes twice as much labour to clean it out. Yet other sows will go backwards to do their business, thus keeping themselves and their bedding as clean as a new pin. A fortune is waiting for someone who can devise a simple method of always getting them to perform backwards instead of forwards. Slats behind the sows will help take the liquid away, but invariably the droppings are left.

17
Teeth Cutting

A pair of good-quality snips are necessary for cutting piglets' teeth properly, without leaving any jagged edges. Five to ten minutes after they are born the piglets will be dry enough to hold. Rest your left leg on a step to enable you to rest the piglet's backside in your groin, and with your left hand hold it around the neck, press in the corner of the mouth on the opposite side from you, which will induce it to open the mouth sufficiently to enable the snips to be inserted. Then snip off the incisor teeth, top and bottom on each side as near to the gum as possible without damaging it. You can make no mistake here, as these will be the only teeth visible. Mark the piglet and put it back; the pigs themselves will make nothing of it, and each one will not take more than fifteen seconds to do.

18
Identification

Identification of your gilts can be done in several ways. If you only have one or two then you will know them by sight and probably give them names. Ten or a dozen or more will need some sort of identification.

Numbers can be tattooed into their ears at about two weeks old, and these will be readable for life if done properly. At the same age, little *V*s or notches can be cut out on the outside of the ear, no deeper than five centimetres, which the pig will hardly feel; again, once done this will always be visible. Enough combinations can be realised in this way for a good-size herd without any other identification.

There is an electric drill type of tattoo machine for tattooing

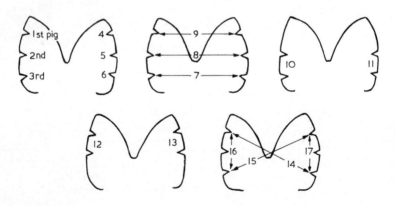

Identification and ear markings: notches cut in piglets' ears for identification when grown to sows

IDENTIFICATION

numbers, best done on the pigs' backsides; very effective once done, this lasts for life.

Dozens of makes of tags are also available on the market; these require a hole cut or punched in the ear to fit. One drawback of the tag is that it can get pulled out and lost, but some of the modern sort, which are inserted in the middle of the ear, take some shifting.

There is a little plier-type tool which will make a small hole in the ear, making identifications similar to the notches on the ears.

19
Castration

It was around 1976 that boar meat took off, with about five per cent of male pigs being slaughtered entire; but now, some five years later, as many as fifty per cent are non-castrates. There is no doubt that entires have an increased growth rate over castrates which could be as high as ten per cent, with a higher percentage of lean meat. The pork-eating public seems to favour this sort of meat.

The argument for the castrates is what it has always been—that the meat is free from what is called the 'taint' in the entires, what we call in my part of the country 'pissy pork'. It is not so noticeable in the taste, but more the smell when it is being cooked. The anticastration people's view is that if the pigs are grown fast enough they are slaughtered before they have a chance to mature and develop this taint. I am sure this is true, but there is often a pig that gets a setback for one reason or another, and could be as much as a month older than the others before it is big enough to go to the abattoir. This is the one that gives the entires (boars) a bad reputation. In general the entire boars are acceptable for the pork trade, but not for bacon or heavy hogs.

The removal of the boar pig's testicles is a simple operation, once you have done one or two. There is a simple gadget whereby the patient is held on its back and has a strap for each leg, which enables the job to be carried out by one person. But I do not recommend this technique as you need clean hands for the actual castrating—which would not be the case if the catching and strapping was done by one person, unless of

CASTRATION

course the hands were washed and dried between each pig.

The best time to castrate is between ten and twenty-one days. The pigs are big enough then for the testicles to be handled, and not too old to give them a setback. It is obviously painful, and they will squeal all the more for being caught in such a position. So it is better to make sure the mother is out of sight and sound.

Have the pig held in one of the positions shown, whichever you prefer (see photograph), up on the chest for the smaller ones and in between the legs for the big ones. Clean the area with methylated or surgical spirit on some cotton wool.

With your second finger, bring up the selected testicle from the groin into its usual position in the scrotum; with your thumb and index finger press together underneath the testicle to reveal it proud above the thumb and index finger so that you feel you have a tight grip on it. With a scalpel blade make the incision nearer the groin, so it will be at the base when the pig stands up, and can drain. Make the cut about half an inch long (depending on how big the testicles are), through the outer skin and the inside membrane, which will probably seem as one, to reveal the testicle which will pop out if your grip is right and the cut is big enough. It will not matter if you cut the testicle in the process. You may now release your grip, as it is unlikely that the testicle will slip back. With your two fingers and thumb, catch hold of the testicle and pull out about one inch, to enable the scalpel blade to be inserted between the white cord and the red one. Sever the nonvascular white cord, leaving the red cord (carrying the blood vessels); pull this cord out further, twist it a couple of times, and sever with a scraping movement up and down. With this action the blood vessels will not bleed as much as with a straight cut through. Some people prefer to pull this cord right out of its roots, but there may be a risk of internal bleeding with this method. Repeat the operation with the second testicle. Antiseptic powder can then be used as a precaution against infection. Put the piglet back into clean straw so as not to get any dust in the wound. The piglets will feel sorry for themselves for a day but will soon forget it. When you get used to castrating, each pig will not take more than thirty seconds.

Ruptures, or Scrotal Hernia
I do not advise the amateur to tackle this problem, and it is best left to a vet. Only the outer skin of the scrotal must be cut; the inner membrane skin must be got out with the testicle inside, and the intestines pushed back and then stitched. The testicle can then be removed. But as more boar pigs are now being bought as pork pigs with little or no price deduction it is not worth the trouble or expense. Navel, or umbilical, hernia is difficult to treat and little can be done about it. Do not keep ruptured pigs longer than you have to; they will generally get by to pork weight without problems.

Ruptures are an hereditary weakness, and sows that are prone to throwing ruptured pigs should not be bred from. However, more often than not the trouble can be rectified next time around if the boar is changed.

But, nevertheless, never buy them knowingly for they are another hazard to cope with.

20
Ringing

If you want your pigs to go out on grass but not root up the field, the only way of preventing this is by ringing. You may get away without ringing sometimes, depending on the attitude of your pigs and perhaps in summer when the ground becomes hard—then they may not bother.

There are two types of ring, one where the ring is put in around the top of the snout, precisely where the pig roots from, with a pair of pig ringers especially made for the purpose. The other sort is a big ring, put in right between the two nostrils. This ring opens up on a hinge at one end of the half circle and a point on the other. The ring is big enough to be held in the fingers and pushed through the flesh between the two nostrils with the point, then it must be clipped shut to complete the circle again. It will hang down, rather like a bulb ring. When the pig tries to root, the ring swings out; thus touching the ground with the end of its nose causes discomfort and persuades it not to root.

Ringing is one of the more unpleasant jobs, with the pig squealing its head off, but it only lasts a few minutes and is soon forgotten. It is not the putting in of the ring that makes the pig squeal so much, but—as with castration—the very fact that it is caught in this position. I would suggest that if you have no experience in ringing, you ask an old farmhand who has done it all before for his help, or ask your vet to do it for you.

The procedure for both types of ringing is the same, requiring a place where the animal can be closely penned, like a weighing machine, a pig stall, an individual feeder or a farrowing crate.

Rope for ringing

Have a piece of thin rope (baler twine is not thick enough), at least six feet long, fold it in half and make a slip knot (see diagram) with the noose in your hand. It is quite easy to get the rope into the mouth, provided the head is not constantly kept down. Just to secure the end of the nose is not good enough. The noose must be further back in the mouth, and then kept tight; pull the rope slightly up and back over her, securing the rope once around an iron bar and holding it in a position that enables the ring to be put in. Do not tie the rope, because in the event of the pig falling down it would be left hanging by the nose.

There is a tool for catching pigs by the mouth, which has a wire noose at one end and what resembles a garden fork handle at the other. In my view this is not the best way to ring pigs (unless they are small), as it takes a very strong man to hold a pig still enough for ringing with any degree of accuracy.

21
Fighting and Mixing Pigs

Pigs can be mixed together quite successfully if one or two important points are adhered to. As a general rule the smaller the pigs the easier it is to mix them. When mixing two groups try to put them in a house that is new to both groups. There will always be some fighting, to see who is going to be the boss—as with most animals, there is a pecking order—but provided the loser has room to run away, the victor will not continue to fight and will bear no malice. Big and little pigs seem to mix well. If you have a constant fighter who is going for everyone in sight, give it a few strokes with a whippy stick. This will do the offender no harm except that it will sting for a while, and it generally does the trick.

Sows are probably the most difficult to mix, but after they have sorted out the pecking order you seldom have any trouble. There are one or two tranquilliser drug injections on the market which subdue the pig for a few hours after mixing.

Sows and litters mix together well, apparently due to the presence of the litters.

Some strains of pig are more aggressive than others, and people try all sorts of means to get them to mix more easily. Mange wash is poured over them. Diesel oil, used tractor oil, linseed oil, even the wife's old or even new perfume is used, but in my experience it makes little difference; if they are going to fight, they are going to fight.

It is asking for trouble to try to mix a pig that is unwell for one reason or another.

FIGHTING AND MIXING PIGS

One or two chains are frequently hung in the pens. This may relieve boredom, but I do not think it will stop pigs fighting. If the whole pen can be let out to run around for a bit, this takes their minds off fighting. A paper sack or two thrown into the pen will help get rid of their excess energy; also, a few logs will keep their attention for a while, and the same procedure can be applied for tail biting just to give them something to think about (see Chapter 12).

Gilts cannot be expected to run with sows, with or without litters. They are fine together out in the field where there is plenty of room, but it is the sleeping arrangements that cause the trouble.

I would never mix pigs of 100lb or more. It is not worth the risk, and they will be culled in a few weeks anyway if they are being reared for meat.

One advantage of farrowing sows in batches is that the piglets can be successfully evened up, if one sow has a large litter and one a small litter. In general, a small number born means large piglets, due to the fact that over the past months of pregnancy they have had all the nourishment that would normally be divided between a larger litter. This factor is the first thing that always crosses my mind when a sow starts to farrow, if the piglets are really huge; I half-expect there to be only a few. If a sow has only five or six large piglets, I take four of the biggest from the sow that has twice that number, and add them to the litter. In some cases, if I have two very uneven litters born at the same time, I give one mother all the big ones and the other mother all the small ones. All pigkeepers strive for a good number of even piglets, but it does not always work out like that, and often there is a small pig.

A sow will produce similar litters each time she farrows, particularly if the same boar is used. If I am not pleased with a litter—either in numbers born, unevenness, or size—I try changing the boar. Piglets born within a few hours of each other or on the same day can be mixed with no trouble. If there is a period of forty-eight hours or more, then I hold the piglets up by their legs and rub them in the dung or urine of the sow that I want to accept them; after a feed they will smell like her piglets anyway.

Sometimes a sow will attack the piglets, at any time from when the first one is born, but hardly ever after she has finished farrowing. I put this down to the fact that giving birth is not the pleasantest sensation in the world and she is bound to be in some pain; she takes it out on the first thing she sees, and that is the piglets. A sow could end up killing a considerable number. In my experience the very action of the sow going for the piglets makes you panic; in fact it is not as bad as you imagine. I have often seen a sow go for a piglet and have it right in her mouth, and right up to that moment she seems as if she will kill it for sure. She then gives the impression of saying, 'Well, I'll let you off this time,' and lets it go.

In these situations don't hesitate—put the piglets in a box under or beside the heater, or if you can block the piglets off in the creep, leave them there at least until the mother is lying down satisfactorily or until she has finished farrowing.

When weaning at not less than five weeks, a common practice is to mix the sows and litters together, known as multisuckling. The advantage of this is the amount of space saved. Four or even six litters may be put together in one house, and one

Applying Stockholm tar to soothe tail biting on the end of the pig's tail

house is easier to clear out than a lot of small ones. *Never multisuckle piglets that are not at least a fortnight old,* as the risk of them being lain on is too great. Try not to mix together piglets that have more than a week's difference in age.

Tail biting is a vice. Stockholm tar is as good as anything to use, as it covers up the affected part of the tail, and the pigs do not like the taste of it. Take a one-pound jar three-quarters full and dip the tail into it.

22
Artificial Rearing

It is sometimes the pigkeeper's misfortune to be in a situation where the piglets must be artificially reared. The sow may have died, or through illness may not be able to supply the piglets with milk. Today sow milk substitutes are much improved, and there is every chance of successfully rearing most, if not all, of the litter. The critical factor is the amount of colostrum (the sow's first milk) the piglets were able to get. If there was twenty-four to forty-eight hours' suckling time, then the piglets should have a good chance of survival, but the shorter the suckling time, to get some of that vital first milk, the harder the task will be to rear them artificially afterwards.

It is not a good idea to bottlefeed piglets; it sounds all right in theory, but it is too time-consuming, and once on the bottle, it is difficult to get piglets to take anything else. Induce them to suck up some sow milk substitute by putting a little in a saucer and holding it up under their noses, getting their mouths wet. Two or three times like this and they will soon taste it and come looking. Never force them, as this will have the opposite effect from what you are trying to achieve. Have just half an inch in a bowl to start with, so that they cannot get their noses in too deep and choke themselves. A small poultry drinker is excellent for this purpose (see diagram); failing this, an old fryingpan combined with a sweet jar or any jar-like vessel will do. Pour the milk into the jar, put the inside of the frying-pan down on top of the jar, then quickly turn it over, thus trapping the milk inside the jar, which can then be regulated by lifting the jar slightly. This will stop the mad scrummaging piglets from

ARTIFICIAL REARING

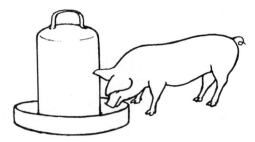

Poultry drinker for rearing piglets taken from the sows

getting into the pan and getting wetter than they should. The jar will have to be held, but only for a half a minute or so, as they should soon drink the substitute all up—hence the ideal poultry drinker.

Alternative drinker for piglets, using a frying pan with a jar or tin

It is a nuisance having to feed piglets six times a day or more, but this period lasts only about a fortnight, after which time they can be weaned onto baby creep pellets and water. Pigs successfully reared on this system often grow faster than the ones with their mothers, having got over the weaning check early, and go steaming ahead. Directions will be given on the packet of milk powder. Abide by these directions to the letter, and *never* overfeed; piglets will not die from hunger, but will die from scour if fed too much.

23
Diseases

This chapter does not list *all* the pig's ailments and diseases, but it does mention the more common troubles and the ones more applicable to the keepers of small pig herds for whom this book is intended. The diseases are grouped under four headings: *notifiable diseases* (page 110); *skin diseases* (page 112); *worms* (page 114); *other diseases* (page 116).

It is particularly pleasant to be able to say that the person with one or two sows or fattening pigs would be deemed unlucky if he or she came across virtually any of the diseases mentioned below, provided that the sows were for the most part outside with a good house to go in and a varied diet of grass, vegetables, meal, etc.

It is the term 'intensive' that pinpoints the cause of all the pig troubles today. Wherever there are large numbers of animals massed together, it is often not very long before disease breaks out. With perhaps a thousand pigs under one roof, all breathing the same air, it is a work of art to keep them on the straight and narrow; on some intensive units the animals are continually stuffed with antibiotics to keep down some disease or other, when all that is needed is a bit of space and fresh air. The hygiene of your pig houses is up to you. Take every opportunity to clean out and disinfect your pens and houses whenever possible.

If your pigs are unfortunate enough to contract any disease, the procedure to eradicate it, in hygiene terms, is always the same. Firstly clean out the pen (steam cleaning is excellent); then scrub with hot soda water (a handful of soda to one gallon of water) and disinfectant. Vary your disinfectants from time to time, as with antibiotics the bugs are apt to become immune to one particular sort if used over a long period.

Rest your pens as much as possible in rotation. Never be in a situation where all your accommodation is full; always have a pen of each kind (farrowing pen, fattening pen) empty in case of emergency and, more important, to break any disease cycle that might be building up.

It makes sense not to have all your pigs under one roof. Try to keep the farrowing pens and weaner pens away from each other, at least not in the same airflow.

While it is of some value to rest a pen sandwiched between two others with the same air circulation, it is not half so effective as resting a house or pen that is completely isolated from the others. Leaving a house empty for three weeks or more is far more effective than all the disinfectants put together.

Notifiable Diseases

Notifiable diseases come under the official control of the Ministry of Agriculture, who must be informed immediately a case of one of the following four diseases is confirmed.

When any of these notifiable diseases is in your area, take sensible precautions: keep people and vehicles away from your unit; or, if they must come, make sure they are adequately disinfected.

Anthrax

Anthrax is a disease we don't have very often in this country. Usually it attacks cattle, sheep and horses, but it has been known to affect pigs, who can catch it from contaminated imported feedstuffs, or by eating unboiled swill containing infected flesh.

There are two forms of anthrax. With the neck and throat form, the pig goes off its food and has a high temperature with a large swelling under its jaw, which affects its breathing; it has difficulty in breathing and froths visibly at the mouth.

The other form of the disease is the abdominal type, where the pig runs a high temperature and may have bloodstained diarrhoea, and will suddenly die. Humans can contract this disease, which makes anthrax extremely dangerous. As with all diseases, or if your pig dies for no apparent reason, contact your vet.

Foot and Mouth Disease
Carcases of infected imported meats which reach our pigs through the swill bin are the primary cause of foot and mouth disease. Anyone feeding swill must be very alert to lameness in their pigs, which will probably be the first signs of this disease.

Initially the virus produces a fever, and blisters will appear on the feet and snout of the pigs. Pigs will be off their food, but within a day or two will seem to recover; it is then that you will notice the lameness caused by infection affecting the blisters and this causes the claw to separate. Call in your vet if you are suspicious.

Animals will survive foot and mouth, as they do in countries abroad, but in Britain the law is to destroy all animals on the farm even if only one animal has the disease confirmed. As with all four of these notifiable diseases, full compensation is payable at the market price.

Swine Fever
Another disease which, thank goodness, has virtually been eradicated from this country, it is contracted through imported pig meat. Pigs with this disease run a very high temperature and will sway about on their hind legs as a result of kidney damage caused by multiple tiny haemorrhages, which occur all over the body; diarrhoea and coughing may be present.

The pigs will go off their food and huddle together and shiver. High losses will occur. There seems to be no effective antibiotic treatment. As with any pig that dies of unknown causes, consult your veterinary surgeon immediately.

Swine Vesicular Disease
This disease in its present form is quite new in Britain and has only recently flared up here. Hitherto this strain had only been reported in Hong Kong and Italy. It is indistinguishable from foot and mouth disease, and indeed when it first appeared it took quite a while to determine its true nature. Fortunately it only affects pigs, but is contracted through untreated swill, as with foot and mouth, and there is no satisfactory treatment.

Skin Diseases

Lice
Lice on pigs are quite common. You will find them on most pig farms, but really there is no excuse for having them as they are easily got rid of.

The pig louse lives only on the pig. The female lays eggs at the base of the hair, and lice eggs will hatch in twelve to twenty days; the larvae, called 'nymphs', will undergo three changes before they become adults. They can easily be seen on the pigs and seem to prefer it around the neck and ears and right along the back. The lice look and walk much like tiny crabs, puncturing the skin and sucking blood. In warmer climates they multiply more rapidly, and naturally they cause the pig to itch and scratch.

Louse powder can be purchased, and if sprinkled directly onto the lice they will all be gone in a matter of hours. The mistake is made by forgetting that there are eggs that will be hatching up to three weeks after the first delousing; so I would recommend a derris louse dressing once a week for three or four weeks. If you are suspicious of your housing try to put some louse powder in a puffer pack and puff it around the house before you put in fresh pigs.

Pig Pox
This disease can be spread by a carrier pig which itself will show no signs or symptoms. It is caused by two viruses said to

be related to the smallpox virus in humans (no danger of humans catching it). Pig pox causes blisters around the udder and teats, and when these blisters burst little ulcers are produced which are quite sore, and are further irritated during suckling. The pig will need to be washed in an antiseptic solution; consult your vet, who will prescribe treatment.

Pityriasis
A skin disease often confused with ringworm, pityriasis occurs only in piglets when they are about three to four weeks old.

The rash usually starts on the belly—red spots with a depressed centre, which form a brown scab—and comes away to form a deeper crater. Before the rash the piglet will be off colour and may vomit and, like a child with chicken pox, when the spots come out fully it feels better. Consult your vet, who will supply you with an antiseptic spray which will clear it up.

Pityriasis is a congenital condition, so try to use a different boar on a sow that has produced infected piglets.

Sarcotic Mange
This is a skin disease but is caused by the mange parasite which is too small to see with the naked eye, and is similar to scabies in humans. The parasite burrows into the skin, or the ear, which is a favourite place to start, where it lays eggs which hatch out in less than a week into larvae; these rapidly change into adults and very quickly cover the whole body, causing the pig to scratch and rub constantly. The pig will soon lose condition. Apparently healthy pigs will carry the parasites, and are a nuisance to others pigs, particularly any that are run down or off colour.

Scrub with a mange wash any pig that may seem out of condition—another reason for giving the sows a mange wash before they are due to farrow. In severe cases, several washes may be necessary, with a few days between each wash, to make sure you kill the newly hatched parasites. Consult your vet, who will have experience of these things and can immediately recommend the right treatment for your particular problem.

As with all professionals, the advice of vets is not cheap, but when weighed against perhaps more vets' bills, extra food to feed sick animals back to health, or even pigs' deaths, the calling in of the vet at the outset could be ten times less expensive in the long run.

Worms

Worms can be a serious problem, but can be easily controlled with the effective modern pig wormers available today. Worms can be responsible for many conditions in the pig; the old saying, 'You must have worms,' when someone could not stop eating, was probably referring to the pig. A pig with a bad infestation of worms may need to eat twice as much food to keep in the same condition as a pig with no worms.

Ascaris

The most common worm in the past, or should I say the most recognised, was the round worm (*ascaris*). These are the worms most visible after worming; they are white and can be up to twelve inches long. They live in the small intestine and can be easily dealt with simply by adding a worming preparation to a feed. When the food passes through the small intestine the worm eats it and dies and comes out in the dung.

Hyostrongylus

The only worm that truly lives in the stomach, *hyostrongylus* is no thicker than thirteen-amp fuse wire, not half an inch long and red in colour. It can cause inflammation of the intestines and can lower the condition of the pig. It is well controlled by a modern wormer.

Metastrongylus

The long worm, not perhaps so common as the others, is a white threadlike parasite, approximately two to three inches in length. These worms are somewhat different to the others in that they need the presence of pasture to complete their life cycle. The female lays her eggs in the lungs. They are coughed up and swallowed and then come out already hatched in the droppings. They are then eaten by earthworms. It is said that an earthworm can carry as many as two thousand larvae, and it is these earthworms if eaten by the pig that start the cycle all over again. Humans, when rather run down, seem to catch everything that is going; it is the same with pigs which are off-colour—they pick up the lung worms.

Earthworms from the pasture can be examined for larvae content in a laboratory to help determine the extent of the problem.

Pigs that are kept in yards and cleaned out regularly with no access to pasture should not be troubled with this worm.

Oesophagostomum
The nodular worm, so called because the larva burrows into the mucus membrane of the intestine causing small nodules, returns to the bowel when mature. The nodules can cause inflammation and diarrhoea and can interfere with the heat periods of the sow.

Trichuris suis
This is the whip worm, so called because it attaches itself to the mucous membrane of the large bowel by a long thread which resembles a whip. The eggs are laid in the large bowel and go out in the dung, where in warm and moist conditions, particularly in summer, they develop into larvae in about three weeks. When eaten by the pig they find their way to the gut wall, burrowing quite deeply. It is this carve-up of the gut wall that causes scouring. It is thought that the holes made in the gut wall increase the likelihood of *E. coli* and other gut troubles.

Worming With many of these parasites and diseases, there is so much that we do not know or understand; we can only guess what is happening and try to take appropriate measures. If you have a worm problem consult your vet, who will advise you on what wormer to use. At the present time there is no single wormer that will kill all types of worms, but there is now a very good subcutaneous injection wormer on the market.

I would suggest worming your sow at least a week before she farrows, which would mean she would be wormed approximately twice a year. The little pigs should be wormed after they are weaned and eating well, at about seven weeks, and again a month to five weeks later; and, of course, don't forget the boar—two or three times a year.

Worming pigs is very easy. Directions, including the right measure, will be provided with the worming drug. There is no need to fast the pigs beforehand, just mix the required amount in the normal feed. I am not too keen on the injection-type wormers; not that they don't do a good job, but it is a nuisance having to sterilise the syringe and needle every time you want to worm a sow, and she will not appreciate having a needle being stuck into her, either.

A new concept in pig worming has just been started by at least one firm of millers in which the worm drug is incorporated in the feedstuff at the mill before you receive it; it is then fed normally to the pigs for a fortnight, and it is claimed that this will kill adult worms, larvae and eggs over this period. I like the idea of this very much, as so often we worm the pigs while forgetting that there are probably eggs that will hatch in a day or two and reinfest them. Initially this service was instituted for the large pig-units to use by the ton, but I see no reason why it could not be mixed up and sold by the bagful for the smaller pigkeepers.

Other Diseases

Atrophic rhinitis
A disease comparatively new to this country, probably originating in America and Scandinavian countries, I would

say the condition was virtually unknown here a few years ago. The Swedish Landrace was blamed for introducing it into this country. Now it has assumed epidemic proportions in some areas and can be very expensive if your pigs have a bad bout of it.

There has been considerable discussion among pigmen as to making this disease a notifiable (see page 110) one, with the view to eradicating it from this country. This has proved to be completely impracticable, as there are many variations and secondary side effects to this comparatively little-known disease.

Sneezing would be the first indication of this disease, particularly after feeding; it occurs because the nasal system is becoming blocked and twisted, resulting in the nose being distorted. By this time the damage is done, and from then on I believe it will take its course. One can medicate to try to relieve, but it is like taking medicine for colds, they never seem to do much good. Their eyes are likely to run and there will be a discharge from the nose, often with blood. With these symptoms the pig will not eat as well as it should, due to the tenderness of the nose, and has difficulty in breathing, thus making it unthrifty. We have had considerable success with antibiotic injections, starting as early as a week old.

Atrophic rhinitis produces secondary infections which are similar to virus pneumonia (page 119), so half the time you do not know what the pigs have; that is why this disease is so difficult to treat and control. Inform your vet, who will inspect and advise accordingly. Rhinitis occurs primarily in piglets up to six or eight weeks old, then they may become carriers of the disease.

Control Medication of food and water is helpful. Antibiotic injections can be given. A vaccine has just come onto the market, but I do not know the value of it. Many pig farmers have become so despondent with this disease and the difficulty of treating it that they clear out the whole herd of pigs and start again after an appropriate time. Let us hope that the new vaccine will be the answer.

Aujeszky's Disease
Not a common disease, this is transferred through carrier pigs. Only young pigs show symptoms, and pigs under six weeks will often die, having shivers and throwing fits, with a high temperature caused by the virus attacking the nervous system. Pigs over fourteen weeks seem to get over it, often just having a twenty-four-hour fever. Aujeszky's disease is sometimes referred to as 'mad itch'. Cattle and sheep as well as cats and dogs are known to contract this disease, but only a few cases have been confirmed in this country. I have not seen this disease myself, and there seems to be a difference of opinion as to whether pigs really do itch or not. There is no treatment for this condition.

There is at this time pressure being put on the government to make Aujeszky's disease a notifiable one (see page 110) in an effort to stamp it out altogether.

Escherichia Coli (E. coli) Troubles
A severe problem in some herds and another of these disorders that we do not fully understand. It is sufficient to say that if nothing is done about it, huge losses of little pigs can occur. There are two main areas of concern: that of piglets when they are first born up to a few weeks old; and after they have been weaned.

There are millions of *E. coli* in the gut at any one time, and it is not quite clear why an upset occurs. There are many different sorts of *E. coli*, and biologists can analyse these and tell you which particular strain is causing the trouble so that the right antibiotic can be used against it.

E. coli causes scour; piglets are sometimes born with it, being thin and bony with big tummies and runny behinds. It is essentially the sow's milk that is at fault in the first instance. Why some sows produce this condition is something of a mystery. It is the first-farrowing gilts that are the worst offenders, presumably because they are not old enough to have built up a resistance against this disorder.

Treatment Sows can be vaccinated against *E. coli* with considerable success. There is now a new idea on the market which is incorporating a dead vaccine into the feed. This has to be done continually as it does not give immunity for very long. First results are encouraging.

A dispenser can be used for the piglets orally, which generally clears up the condition. A few squirts of sugar water in the mouth will sometimes help them, if they are very bad and dehydrating. Injections can also be used.

Inside the gut wall there are what can be described as hairs or fingers which sieve and use the food passing through them; these get damaged or eaten away by the *E. coli* and this causes the food to go straight through the pig. Consult your veterinary surgeon.

Enzootic Pneumonia (Virus Pneumonia)
This is another of those conditions that many pig farms have lived with, as once this virus occurs it will always be present in the herd. The only way to eliminate it is to clear out your entire stock and start again after about a month.

This disease is not usually a killer but can be a costly business. It is an environmental problem and is magnified by cold draughty housing and too much variation in temperature. Virus pneumonia is caused by a minute germ which constricts the bronchial tubes and makes the pig cough, which damages the lung tissue. It is the secondary bacterial pneumonia that causes the trouble. The pig will be reluctant to do anything, will not want to eat, but will not cough too much unless it is made to move around. If caught in good time the pig will get over it in a few days, but if left, will take many weeks to recover.

A sow that has farrowed and is finding it hard to maintain her condition is more likely to contract this virus. New stock can introduce enzootic pneumonia into a herd and cause havoc, or new stock can catch it from carriers in the herd and cause it to flare up again.

Erysipelas
Piggeries which are susceptible to this disease often vaccinate their stock regularly. Everyone should vaccinate all breeding stock from as early as five weeks, and every six months thereafter. The vaccine is cheap and easy to administer as one injection.

Pigs of all ages are vulnerable, but pork and heavier pigs are most at risk. Patchy skin and lameness are good pointers, with very high temperatures which cause constipation. The victims will lie about, be reluctant to walk and appear stiff-jointed.

The erysipelas bug can live in the ground and is most prevalent in summer, in hot, muggy conditions. A pig will sometimes die within twenty-four hours with the loss of appetite the only symptom, but this is the exception rather than the rule. Call in your vet as soon as possible. If caught early enough, penicillin will provide a complete cure.

Joint Ill
Joint ill is a real problem in some herds, and pigkeepers have great difficulty in stamping it out. Joint ill, so called because the leg joints of two- or three-week-old piglets swell up, is caused by a common germ gaining entrance to the pig's system via the navel cord, in the first few days of life; or through facial scratches caused by fighting (another reason for cutting their teeth when born); or sometimes through raw knees due to fighting for the teats, if the floor is rough or pitted. The disease is very painful for the little pigs, so much so that they will just lie about, with a temperature, not even getting up to feed, therefore quickly losing condition. If treated early enough, joint ill responds well to antibiotics. It is difficult to know the exact cause; as with many of these troubles, we do not really know enough about it.

Prevention Spray the navel cord once a day with an anti-

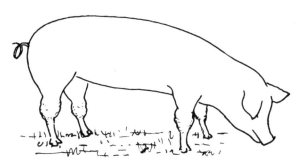

Joint ill causes very painful swollen joints

biotic spray for the first three days. Give the farrowing quarters a good scrub and disinfect well between farrowings. If the floor is rough and pitted, put a thin skim of equal parts cement and sand over it. A weeping abscess on the sow's udder could be the cause: treat with antibiotic spray.

Lameness
Lameness in pigs is caused by a number of factors, in young pigs it could be erysipelas, joint ill, or occasionally rheumatism or arthritis.

In sows, lameness is a common complaint; often the feet get damaged in the first instance, and after a time infection sets in. If you see a sow holding up a leg with a damaged foot you will know that it must be very painful; she will quickly go back (lose condition) if nothing is done about it. Often when sows get foot problems it is an old wound which has never healed completely; the least knock starts the trouble off again. Antibiotic injections will help and a kaolin poultice is good but not always practicable.

Inspect the pigs' feet whenever you get a chance. I find a good place for this is when they are in a farrowing crate, lying down, where one can quickly trim them up if necessary with a good pair of feet clippers or secateurs.

If foot troubles persist, look for the cause—this could be uneven surfaces, too rough concrete, or continually standing in mud, etc. If rheumatism or arthritis or general stiffness is apparent, look for faults in the sleeping quarters, damp and cold floors, or draughts (see Chapter 4). Boars too are vulnerable to lameness, mainly due to their occupation, running around serving the sows.

Mastitis

Mastitis in the sow's udder is a major factor in piglet losses; they are literally starved to death. The first indication of this will be the squealing and unsatisfied piglets, not lying down after a feed but continually trying to suck the mother. Her udder will be hard, and milk will be nonexistent, watery or clotted. She will be reluctant to let the piglets suck because of the soreness of her udder, and she will run a high temperature. This condition can be caused by draughts in the farrowing house or by germs entering a damaged tear or cut. Send for the vet quickly; she will respond well to penicillin injection.

Scour

Scour, diarrhoea, 'the runs', call it what you will, is a condition that no one likes to see, as it will be doing the pigs no good whatsoever.

It may have a number of causes. *E. coli* could be causing it when piglets are first born; or there is what is known as the three-week scour, which coincides with the sow's three-week cycle, sometimes upsetting the milk supply, and gives the piglets a white, creamy looking scour. Usually this is nothing to worry about and after a day or two will disappear as quickly as it came. In these situations I find reducing the sow's food helps, which in return reduces the milk supply. As with all scours it is nearly always beneficial to reduce the food. Piglets will sometimes scour when still with the sow, due to eating too much creep feed too quickly, before their stomach has had time to adjust.

Another critical time for scour is after weaning, when the piglets no longer have their mother's milk to sustain them. If a particular batch are troublesome, I often give them just barley meal with a little milk powder for a day or two before going on with the growers' diet.

Nearly all scours can be attributed to *E. coli* in one form or another. Another reason for scour is if mouldy straw or food is eaten.

Medication can be given in the food, or perhaps best in the water supply. There are many more conditions that make pigs scour, but rather than confuse the issue, I think it is best to say see your vet.

Splay Leg
This is a condition found in piglets of Landrace parentage. It becomes evident a few hours after they are born that one or more of the litter have splay legs, that is to say, their back legs go forward or out sideways when endeavouring to stand; occasionally it is the front legs too. If the pig can suckle, full recovery can be a matter of days. If helped to feed there is a ready response.

In my experience splay leg occurs only very occasionally in a Landrace strain. When we get one it seems to me that a little harness could be invented which would keep its legs in place for a day or two.

Splay leg piglets are more prone to being lain on, as they do not bother to go back under the heater light (see page 68) to lie down. Bedding will help them keep their feet.

24
Poisons

It is very rare for pigs to be poisoned, even though they are apt to eat everything in sight. However, it is as well to have some knowledge of the substances they might eat and be poisoned by.

Acorns
Acorns are loved by pigs, but the shells have a high tannic acid content, which under certain circumstances can cause gastro-enteritis and make pregnant sows abort. Usually they will drop the shells and just eat the kernels. If poisoning is suspected try to get them to take some castor oil or linseed oil mixed up in some meal and milk.

Bracken
This is known to be poisonous, the rhizomes (roots) more so than the stems and fronds. If the pigs are being fed with nuts or meal and have an adequate water supply they should come to no harm. Bracken has been known to cause dead piglets and mummified foetuses.

Buttercup
If too many buttercup leaves are eaten they may cause soreness of the mouth, with blisters and perhaps diarrhoea. Mix meal with water to make a gruel, as a soothing drink. This is not a common complaint; a few buttercup leaves mixed in with grass when eaten will do no harm.

Coal Tar, Pitch and Creosote
These can be dangerous, causing loss of appetite and weakness.

POISONS

Fodder Beet
The tops, or crowns, are thought to contain poison, but if fed with the whole beet, pigs should come to no harm. Symptoms are a watery scour and nervous and restless behaviour; the appetite remains normal.

Hemlock
All parts of this plant can be poisonous, giving the pig a drunken appearance. Convulsions and fits may be apparent and can be fatal.

Horse Tail
Horse tail grows in damp or wet places. Affected pigs may scour and lose condition. Castor oil or linseed oil mixed with meal and water as a gruel is a good antidote. The plant is difficult to eradicate by spraying, perhaps due to the feathery condition of the leaves.

Laburnum
Laburnum seeds are poisonous to humans; both the seeds and leaves are poisonous if eaten by the pig. Convulsions and death may occur. There is not much of an antidote, but tannic acid can be tried.

Monk's Hood or Granny's Nightcap
Another water-loving plant found by streams and rivers. Poisoning is usually fatal. Death is due to asphyxia. Try tannic acid (15 to 30 grains).

Mould
Anything mouldy if eaten by pigs or any other animal will do them nothing but harm. Mouldy straw is useless as food; perhaps the only thing it could be used for is bedding for bullocks, provided you are sure they are not eating their bedding at the time.

When buying in straw, totally reject any that has the slightest suspicion of mould, which is due to the straw being baled up when wet, when still green, or with too much green grass in with it. Farmers are often in too much of a hurry to bale it up; if they left it for just a couple more days after the combine to quail, there would not be half so much mouldy straw about.

Mouldy barley meal, pellets, nuts, etc should all be disregarded. Throw them on top of the dung heap and cover them up in case the hens get at them.

One can always smell any material that is mouldy, and if very bad it will have a grey mould all over it. One of the symptoms of mould poisoning is pigs off their food; there may also be intoxication and scouring.

Nightshades
There are three types of nightshade—black, deadly and woody; all are poisonous to pigs. The affected animal may become unconscious. As an antidote give thirty grains of tannic acid. Dig up and destroy the plants before the seed disperses.

Ragwort
A plant most of us know, it is seen growing on waste ground and grassland in summer. It has clusters of daisylike yellow flowers. It is common throughout Britain and indeed throughout most of the world. All parts of the plant are poisonous, and still active when dried in hay. It is a tough plant, and animals will not readily eat it. The poisoning effect is accumulative, and symptoms may sometimes not be seen until several days after the stock has stopped eating the weed. Ragwort does not like being disturbed, and continual cutting will suppress it; several applications of spray will eradicate it. Ploughing and rotation of crops is the best answer. Old farmhands will point it out to you if you are not sure exactly what it looks like.

General symptoms are loss of appetite and condition, fast pulse, rapid respiration and weakness; sometimes jaundice may occur.

POISONS

Rat Poison
Some rat poisons are harmful to pigs. Always be wary of them, never put down where pigs can eat.

Rhododendrons
These plants are poisonous to some animals, but pigs seem highly resistant. If poisoning does occur, heavy breathing, weakness and staggering about will be evident.

Salt
Too much salt can be a killer in pigs. There have been instances when feed merchants have by mistake put in too much salt, with disastrous results. It is known that the wet-fed pig can take far larger quantities of salt than the dry-fed one, with no apparent harm.

An unacceptable amount of salt and brine, which is more than salt, is sometimes found in swill due to pickles, etc.

Some animals may die of excess salt, others becoming very thirsty and passing large quantities of water; they may appear weak, and vomiting may occur. Make sure they have a plentiful supply of clean water, and give them a thin meal gruel.

Solanine
Green and sprouting potatoes, when fed raw to pigs, can cause solanine poisoning, more to the younger ones. If the potatoes are boiled it renders them harmless, but throw away the water where the solanine poison is left. Diarrhoea, loss of appetite and condition, and a low temperature are the symptoms. I have given my sows a small steady supply of raw scruff potatoes for years with no ill effects. Too many potatoes at one time will give them diarrhoea.

Water Dropwort
This plant is found in damp places, low-lying meadows and marshes. Although often not fatal when eaten, it will give a nasty stomach-ache and convulsions. The roots resemble a

large dahlia tuber, and are more harmful than the leaves. Castor oil or linseed oil mixed with meal and water as a gruel will help.

Yew

Yew is probably the most common form of poisoning to all animals. There is no known antidote. It is the leaves that are poisonous; the berries, surprisingly enough, are not. Churchyards and large old house gardens are the most common places to find yews.